"*The Titanium Economy* formally recognizes a sector that has long punched above its weight—and one that should be at the vital core of the US economy for years to come. This book should be required reading for anyone passionate about the future of manufacturing. An excellent roadmap for how to win in the twenty-first century."

—Blake Moret, chairman and CEO,
Rockwell Automation

"Industrial technology is crucial to sustaining American innovation and competitiveness. *The Titanium Economy* provides a roadmap for how to breathe new life into the sector and assure its growth long term."

—Cindy Niekamp, board member and
former senior vice president, PPG Industries

"We all know about America's high-tech giants. This fascinating book shines light on another group of innovative, technology-dense companies essential for our prosperity. The fact that these 'Titanium' companies are smaller doesn't make them any less important. Read on to learn about one of the country's great economic strengths, and how to make it even stronger."

—Andrew McAfee, MIT scientist, author of
The Geek Way, and coauthor of *The Second Machine Age*

The Titanium Economy

How Industrial Technology
Can Create a Better, Faster,
Stronger America

The Titanium Economy

Asutosh Padhi,
Gaurav Batra,
and Nick Santhanam

New York

PublicAffairs
Hachette Book Group
1290 Avenue of the Americas, New York, NY 10104
www.publicaffairsbooks.com
@Public_Affairs

Printed in the United States of America

First Edition: October 2022

Published by PublicAffairs, an imprint of Perseus Books, LLC, a subsidiary of Hachette Book Group, Inc. The PublicAffairs name and logo is a trademark of the Hachette Book Group.

The Hachette Speakers Bureau provides a wide range of authors for speaking events. To find out more, go to www.hachettespeakersbureau.com or call (866) 376-6591.

The publisher is not responsible for websites (or their content) that are not owned by the publisher.

Print book interior design by Amy Quinn.

Library of Congress Cataloging-in-Publication Data
Names: Padhi, Asutosh, author. | Batra, Gaurav, author. | Santhanam, Nick, author.
Title: The titanium economy : how industrial technology can create a better, faster, stronger America / Asutosh Padhi, Gaurav Batra, and Nick Santhanam.
Description: New York : PublicAffairs, [2022] | Includes bibliographical references.
Identifiers: LCCN 2022013121 | ISBN 9781541701878 (hardcover) | ISBN 9781541701892 (ebook)
Subjects: LCSH: Manufacturing industries—Technological innovations—United States. | Industries—Technological innovations—United States. | Economic development—United States.
Classification: LCC HD9725 .P33 2022 | DDC 338.0973—dc23/eng/20220526
LC record available at https://lccn.loc.gov/2022013121

ISBNs: 9781541701878 (hardcover), 9781541701892 (ebook)

LSC-C

Printing 1, 2022

To the US industrial sector,
the bedrock of the American future

Contents

The Titanium Economy

Chapter 1

Underappreciated, Undervalued, and Misunderstood

A funny thing happened on the way to the collapse of America's manufacturing sector: it actually didn't.

While innumerable headlines of the past three to four decades have chronicled the decline of US manufacturing, the coverage has almost entirely overlooked a resurgence that's well underway. Much of the industrial sector is thriving, brimming with innovation, offering high-quality jobs with good pay and benefits, and producing strong financial returns. In fact, the returns boasted by some of the companies at the heart of the most dynamic part of the manufacturing sector, which we call the Titanium Economy, have exceeded those of the vaunted

Silicon Valley FAANG tech stars—Facebook, Apple, Amazon, Netflix, and Google.

We lovingly refer to this collection of high-tech manufacturing firms as the "Titanium Economy" because their namesake metal shares a great many characteristics with the companies we identified. Titanium is extremely durable and corrosion-resistant. It is incredibly strong relative to its weight. Perhaps most importantly, many of us likely wouldn't recognize titanium by appearance—but it's around us everywhere we look—in our cars, our mobile phones, our jewelry, sports equipment, surgical tools, and more. These "Titanium Economy" companies are quite similar to their namesake metal—many have persisted for decades, weathering numerous economic storms yet emerging stronger.

The Titanium Economy is composed of a robust collection of firms you've probably never heard of, innovating in ways you've likely not read about, making groundbreaking products critical to trends as leading-edge as e-commerce, electric vehicles, and autonomous driving. What all these companies have in common is a relentless drive for innovation and a zeal to embrace and deploy new technology that rivals the appetites of the Apples and Googles of the world. Some are phoenixes that have risen from the ashes of the old industrial model, having adopted leading-edge technology, while others have been steadily achieving stellar growth for decades. Why haven't they received more coverage? It's because most people, even in investing circles, simply don't know about them.

You Don't Appreciate What You Cannot See

Titanium Economy companies aren't business-to-consumer brands that take out Super Bowl ads, nor do many of them make products that consumers can purchase. The products they do make, however, enable the creation of products and services that are critical to our daily lives, as well as to finding solutions to the many pressing problems the United States is facing, from battling climate change, to creating a more sustainable and reliable food system, to restoring and upgrading infrastructure. Their CEOs aren't often featured on CNBC or the front page of the *Wall Street Journal*, but they're an incredibly driven bunch who have helmed their companies across decades of consistent innovations and profits. And while these companies aren't unicorns, planning headline-grabbing IPOs or hoping for multi-billion-dollar valuations, they are as crucial to the current US economy—and we would argue even more so to our collective future—as the vaunted tech darlings.

For one quick example, consider Qorvo, a little-known company located not far from Disney World in Florida. Qorvo manufactures a critical part for mobile phones and is the only facility in the world making it at scale. Without that part, you wouldn't be able to talk and text at the same time. Nobody outside of the mobile phone manufacturing industry appreciates what Qorvo does, but if that factory stopped production for some reason, many billions of dollars would go along with it.

Wherever we go, and whatever we do, an invisible network of Titanium Economy companies is always at work, supporting our lives in vital but often imperceptible ways. But that invisibility means the brilliance and importance of their contributions have gone mostly unsung. To give you a taste of how seamlessly the innovations of the Titanium Economy are woven into our lives, consider how they might make one day in your busy life a whole lot smoother and more productive.

Your smartphone trills at 6:30 a.m. Before showering, you slip out the patio door for a moment of solitude on your back deck before diving into what is sure to be another hectic day. You're heading out of town in a few hours for a series of critical business meetings, and you're due back tomorrow to see your daughter in her school production of *High School Musical.* If all goes as planned, you'll be in your seat before the curtain rises.

You reach for the new shirt you ordered just two days ago, with expedited shipping, to replace the one you accidentally dumped half an oversized Reuben sandwich on last week. The thought of that Reuben makes your stomach grumble, and you pull up an app on your phone to order scrambled eggs and a breakfast sandwich for pickup on the way to the airport. On second thought, for the sake of the new shirt, you change that to a plain bagel, along with your usual coffee. The last thing you hear as you bolt out the door is your daughter calling out, "See you at the show."

Backing out of your driveway, you notice that your neighbor's roof, damaged by a sudden hailstorm last week, looks brand-new. Even though your home escaped harm, you make a mental note to ask how they managed such speedy repairs.

After stopping to pick up your coffee and bagel, you head for the airport. You're delighted to find that after making your way through a long security line, your breakfast is still nice and warm when you sit down to eat. A quick, uneventful flight later, you use your smartphone to pick up a rental car with no wait, and you arrive at the meeting site with fifteen minutes to spare.

For a moment, you pause in the quiet confines of your car to reflect on how smoothly your day has gone so far before heading to your first meeting.

Now, let's rewind the day. As you stood on your deck at sunrise, you took pleasure not only in the early morning silence but also in the knowledge that the company that made the "wood" for your 300-square-foot deck, a composite made from 80,000 recycled plastic bags, is helping to save the oceans. By reusing bags that would otherwise become waste to construct building material, the company, Trex, saves 1.5 billion plastic bags from clogging landfills and waterways every year.

Along with material innovations, Trex is an innovator in manufacturing. The company's IT team recently made the transition from a central data center to a cloud computing system that links a growing constellation of factories. The move is part of a massive migration of other industrial sector manufacturers from the model of a giant central factory to a more dynamic one of plants situated closer to the supply chain and staffed with a combination of permanent and contract workers. Spending on cloud computing shot up 23 percent in 2021, increasing from $270 billion in 2020 to $332 billion, facilitating vast improvements in operations.

You received your new shirt so fast due to such computing innovation. As soon as you hit the buy button, an algorithm went to work ensuring the shirt could be delivered cost effectively on schedule by, in part, combining your package with other shipments to your zip code. Such efficiencies will become increasingly commonplace given the exponential increase in the computing power of microchips, courtesy of innovation in precision manufacturing driven by companies like NXP Semiconductors. As up to 5,000 supercomputers are expected to come online by 2030, marshaling in the game-changing power of quantum computing, the resulting tsunami of computing power will propel the incorporation of machine learning in every industry, from retail to finance to pharmaceuticals, and is anticipated to deliver more than a trillion-dollar total value to companies by the mid-2030s.

As for your neighbor's roof, it was repaired so quickly because the roofing company, through artificial intelligence, tracks the meteorological path of inclement weather, and with the help of satellite imagery from areas damaged by storms, it can send a message to clients who are contractors within hours asking permission to contact homeowners on their behalf. These advancements are powered by best-in-class data analytics and digital tools that are not only deployed by roofing distributors like Beacon but also the entire industrial sector.

Even your simple breakfast order is part of this new economy. Your bagel and coffee were prepared within a matter of seconds, and tasted so fresh and delicious, based on commercial kitchen innovations driven by companies like Middleby and Welbilt, not to mention that your bagel was kept warm by a temperature-regulating wrapper made from one of many advanced materials being engineered to deliver highly specific properties. Many of them also enhance our efforts to create a healthier and more sustainable world, such as biodegradable materials made by companies like Sealed Air as replacements for plastic. The cup that kept your coffee piping hot is another example. Made of paper rather than polystyrene foam, it's lined with a special coating inside that insulates on par with foam but without the severe environmental drawbacks.

Let's talk about your plane ride back home the next day, which got you to your daughter's performance right on time. It was snowing outside, yet you were able to board your plane thanks to ground support provided by companies like John Bean Technologies (JBT). While waiting for boarding to complete, the temperature on the aircraft stayed comfortable thanks to preconditioned air systems produced by companies like Dabico. And your flight itself would not have been as fast, efficient, and safe without essential parts made through innovations in 3-D printing. Previously seen by many as a tool of DIY makers, 3-D printing has greatly advanced and is being used in countless industrial applications. In 2020, 2.1 million 3-D printers shipped in a market that is predicted to grow 20–30 percent a year through 2028. More industrials have jumped on this technology revolution, most notably in the aerospace industry. When it comes to designing planes, reducing weight is the name of the game. With 3-D printing, a valve that once required several parts can be made into one solid piece at a fraction of the weight.

As you settled into your seat in the darkened auditorium of your daughter's school, you could thank industrial technology for the crystal-clear pitch of her voice as she sang "Start of Something New." Her microphone was manufactured by robots and delivers sound quality once reserved for high-end audiophiles and concert halls.

Throughout manufacturing, companies of all sizes have been driving groundbreaking innovations to make our lives easier, more efficient, or both, every second of the day, from your daughter's microphone to your rental car. They also kept production lines humming during the recent historic Covid-related labor disruptions, delivering healthy returns for their shareholders and investors while at the same time bringing prosperity to their employees and the communities they live in. They have withstood recessions and remained robust franchises, all without any of the fanfare afforded to their tech counterparts. What's more, geopolitical changes, Covid-19, and technologies like advanced computing algorithms, artificial intelligence, robotics, and automation are set to disrupt virtually every corner of industry, accelerating innovation. Yet an overall lack of appreciation of the resilience, strength, and future potential of the Titanium Economy threatens to hold back its growth.

The Titanium Economy is the secret weapon of American industrial revival—the key to ensuring the country's economic vitality as the Fourth Industrial Revolution progresses and the United States faces steep competition from global rivals. The next few years will be critical, as the future growth of the Titanium Economy sector in the United States is far from assured.

Investors, policy makers, and the public at large must appreciate the importance of providing more robust investment in these companies, as well as how their growth brings so many positive ripple effects for individuals and communities, providing more high-quality jobs and boosting the economic prosperity of communities and whole regions. At a time when investors and the media are obsessed with all things digital, we make the case that the stronger engine of growth in the twenty-first century will be manufacturing advances. The digital revolution by no means displaced manufacturing; making things, it turns out, is still the key building block for wealth creation and social improvements.

As engineers—Asutosh and Gaurav are mechanical engineers and Nick is a chemical engineer—we're problem-solvers at heart, deriving a special joy in taking theories and making them practical. That is precisely what drew us to the industrial sector in the first place, and why we've teamed up on this book. We want to solve the limited understanding of the societal benefits that can be derived from the Titanium Economy.

The three of us have spent much of our careers studying the companies in the Titanium Economy, eventually finding our way to McKinsey & Co., and developing a passion for serving them, though each of us arrived at this work through distinct routes, and unexpectedly.

We all came of age as professionals during the dot-com boom, when Silicon Valley called to each of us in different ways. Nick's first job interview was with what was then a small, unknown company called AOL. The company hadn't begun offering commercial service yet, and during his interview, Nick had to ask, "What do you guys do?" He remembers telling his mother afterward, "Why would anyone want to send an email? I like going to my mailbox to get the mail." To be clear, this was in 1993—pre-email days. Wanting to work with products he could see and feel, Nick settled in at a company that made catalytic converters for cars. After arriving at McKinsey, he moved into working with semiconductor manufacturers. He'd made his way to the tech sector, but with a type of tech you could touch.

Asutosh arrived in Silicon Valley in 2000, a time when everything was described as innovation this or that. You couldn't pick up a newspaper or magazine without reading hype about the endless innovation at hand. From the coverage, you'd think all of it was happening among a handful of tech companies concentrated within a 50-mile radius. But he knew better. Asutosh had come to Northern California after a stint in Cleveland, where he'd worked for a series of Midwest-based industrial companies. There, people didn't talk much about innovation, but he saw them doing it every day. To say it made a lasting impression on him would be an understatement.

Gaurav's first journey to Silicon Valley ended before it began. In 2001, he'd been admitted to Stanford University, set to start in 2002, but after the terrorist attacks on 9/11, all student visas from India were put on hold, along with his dreams of making his mark in America. Instead, he took a job at Unilever as a management trainee, and though he'd never expected to work in a factory, he was sent to work in a plant in a rural Indian coastal town. There, he learned the significance of the industrial sector in short order. The whole town, with a population of about 20,000, was supported by the plant, and whenever it shut down for maintenance, even just for scheduled weekend upgrades, Gaurav sensed people becoming anxious. It was a lesson in the real-world impact of industrial manufacturing that he carries with him to this day.

Once we arrived at McKinsey, we began working with industrial-tech companies and became fascinated by their work and penchant for constant innovation. We weren't the only ones won over. Though working in the sector wasn't at the top of mind for most recruits to the company—who generally had their sights set on working in banking, retail, or with the coastal tech firms—we saw again and again that new employees assigned to work with manufacturing firms were becoming just as energized by the work as we were. Their intrigue after learning that they were heading to a company they'd never heard of based in Simpsonville, South Carolina, or Rutland, Vermont, quickly turned to excitement once they were on the premises and their eyes and minds were opened to the remarkable operations of this dynamic, hidden economy and the importance of the work these firms do.

At times we have felt, as one client observed, that we are in the industrial basement while the high-tech party goes on in the penthouse. We're not big partiers anyway, and you know what? We prefer to immerse ourselves in the work going on in the basement because it serves as such critical supporting structure for the rest of the economy. And hey, the leaders of the Titanium Economy aren't letting a lack of adulation get in the way of the great things they know they can accomplish.

Once you begin delving into the ingenious creations of the Titanium Economy, you can't help but be impressed by how they're invisibly enhancing our lives and generating so many positive economic ripple effects. These stories exemplify how the Titanium Economy creates value that is eminently practical, and lucrative, despite being invisible. But it's time to bring the marvels of this economy out into the open. It's important not only that investors and policy makers understand its current vibrancy and potential for a great deal more growth but also that job seekers are drawn to these firms. The companies have many more jobs open than they can fill. And filling those jobs, along with receiving more investment and support through government programs, is crucial to the sector realizing its full potential. That, in turn, is vital to the future of the country, for many reasons.

The United States Is Losing Ground to Competitors

The United States is in a race with many countries to win in the industrial space.

China, accordingly, invests up to 200 times the level of US funding to provide grants and support to a large and emerging network of research centers. And Germany's Fraunhofer Institutes, supported in part by the government, comprises seventy-six technical centers that are focused on applied innovation. Meanwhile, South Korea boasts the highest share of researchers moving between academia and industry, stimulating industry innovation. And for small and medium-sized enterprises intent on becoming smart factories, Korea's AI Manufacturing Platform (known as KAMP), a partnership between a state research institute and the government, shares best practices, accelerating productivity improvements. Finally, in Singapore, the effort to rebrand vocational education with the slogan "hands-on, minds-on, hearts-on" has proved successful at bringing in young recruits. Today, nine out of ten students in the bottom quartile of Singaporean high schools graduate from a technical institute and secure well-paying jobs, whereas in the United States, students in the bottom quartile are at risk of dropping out of high school in higher numbers than other groups.

Meanwhile, the United States is not keeping pace in sponsoring university–federal laboratories and partnerships. The United States needs a robust national strategy for advancing industrial tech, beginning with nurturing stronger collaboration between academia and industry, as these other countries are doing. While the Obama administration created Manufacturing USA (MUSA) to foster more collaboration among government, industry, and academia for advanced technologies, MUSA includes only sixteen institutes—each focusing on a particular advanced technology, such as robotics—whereas China aims to have forty such institutes up and running within the next five years. In addition, MUSA funding levels are low compared to other leading manufacturing countries. Public sector funding for MUSA is typically around $70 million to $110 million annually, which industry sources match 1 to 1. In contrast, Germany provides almost $1 billion in public funding for the Fraunhofer Institute, which combined with industry partner research fees reaches $3.3 billion annually.

With so much international focus on support for industrial-tech innovation, the sector is set to explode, and America must not only keep pace but lead. American economic growth in the twenty-first century must be manufacturing led. The attention paid to Silicon Valley innovation, and the money invested in it, are disproportionate, neglecting the current and potential future value of the industrial-tech sector. There's no question that the digital revolution has been an extraordinary engine of innovation and growth, but making and moving things is still the foundation of a thriving economy, as the supply chain disruptions of the Covid-19 pandemic have shown.

FAANG-Like Returns

The Titanium Economy is not only underappreciated, it is undervalued. Our readers, some of whom no doubt invest in tech stocks, may want to check in with their brokers about industrial-tech companies. Nearly half the companies we studied for this book outperformed the S&P Index from 2015 to 2020, and multiple industrial companies have achieved returns that rival the five companies that make up the tech industry's celebrated FAANG (Facebook, Apple, Amazon, Netflix, and Google). For example, the share price for Trex, the company that uses plastic bags to make decking, increased nearly 5,000 percent over the last ten years, outperforming all FAANG companies and the S&P. Trex's total shareholder returns grew 39 percent compound annual growth rate (CAGR) over the past decade compared to top FAANG performer Netflix's 36 percent. And

$1,000 invested at the turn of the millennium in the Titanium Economy company Middleby, which we'll discuss later on in the book, would have reaped almost $100,000 by 2015—far better than $1,000 invested in Apple, Google, or any other marquee-name tech stock during the same period.

Some 688 publicly traded US companies compete in the emerging Titanium Economy, with five times that number in the private sector. Many of these are scrappy underdogs that know how to compete in tough domestic and global markets from corporate headquarters on American soil. About 80 percent of the public companies are small to mid-cap, with sales ranging from $1 billion to $10 billion and from 2,000 to 20,000 employees. Many of the private companies are family- or private equity–owned, with industrial companies constituting 38 percent of the top 1,000 private companies ranked by revenue across all sectors. Reinforcing the sector's significance in the US economy, these 380 private industrial companies have a combined annual revenue of about $250 billion and posted a compound annual revenue growth rate (CAGR) of 4.2 percent in the five-year time period from 2013 to 2018—outpacing revenue growth of S&P 500 companies, which came in at an average of 2.9 percent.

Further underscoring the strength of the Titanium Economy sector, our research found that an estimated 90 percent of its firms are profitable, with nine of ten firms we studied earning a respectable annual return on invested capital of more than 6 percent. This means that companies don't have to be on a constant spending or borrowing treadmill. In fact, some 85 percent of industrial companies have a relatively low capital expenditure and R&D model. Nonetheless, Wall Street largely ignores the sector, as Steve Miller, CEO of Bulk Handling Systems, expressed. "We are in play, we have positive EBITDA, and would be worth more money," he said, "but some banks still tell us, 'No you're worth less than tech companies.'" The firm he runs uses leading-edge AI and robotics equipment for recycling, making the sorting of trash highly efficient and profitable. He shared his frustration that his company doesn't receive the same credit as Silicon Valley firms applying robotics. "We'd be better off as a tech startup, even without proven products."

Indeed, industrial-tech companies have not received the credit they deserve as relentless innovators. Yes, the Silicon Valley tech companies created unimaginable products like cell phones and search engines, but the industrial-tech sector has embraced and deployed new technology with a zeal that rivals the technological appetites of the Apples and Googles.

In short, the smart money needs to turn its attention to this core sector because the companies in the Titanium Economy have produced stellar results without losing buckets of money, and they've built solid business models, making lasting arrangements with other industrials and critical suppliers, including local universities and colleges for training and education and, in turn, a pipeline of new employees. They're so successful that they need even more employees, however, than that pipeline is yet providing.

FACTORY WORK TODAY IS NOT WHAT YOU THINK

Like the middle sibling, the Titanium Economy is tragically misunderstood, still seen through the lens of the retrenchment of the 1980s and its repercussions. When many plants went dark, entire communities were devastated as long-employed workers were laid off and young people fled due to a lack of jobs, many of them seeking white-collar service work through higher education. The hollowing out made a profound impression on the national conscience, and we're still suffering from an impression formed that there was no hope for an American manufacturing recovery. The public certainly can't be blamed for the lack of appreciation of just how robust the industrial sector's comeback has already been. While the media, pundits, and analysts detail every morsel of news that trails in the wake of software and social media titans, they've almost entirely ignored the story of how resilient, and now resurgent, the industrial-tech economy is and how it is healing communities, beating analysts' expectations, and creating jobs with salaries that far surpass those of many workers in the services and retail sectors of the US economy.

Because of that lack of coverage, the image of a factory that most Americans conjure in their minds is badly outdated, like the horribly loud and dangerous shop floor in the 1979 film *Norma Rae*, with sweaty workers shouting over the din of dirty, hulking machines or hunching over assembly lines, their faces streaked with grease. The historic abuses of the meatpacking, garment, and chemical-processing factories still loom large in the American conscience. But the nature of factories has changed, with strong workplace protections and high-tech equipment transforming them into generally clean, safe, and extraordinarily productive operations.

The wide range of industrial work types is also unappreciated, with most Americans thinking primarily of a few major industries, such as mining, automobile manufacturing, farming, and aerospace, as representative. That's because those industries, and particularly the difficulties they face, get the lion's share of media attention. During the crash of 2008, for example, only the automobile industry garnered any meaningful public attention even though much of the rest of the vast and diverse industrial sector was also struggling.

The emphasis in coverage on such shocks to the industrial sector has meant that the last couple of generations, from millennials to Gen Z, have little awareness that manufacturing still can provide access to a middle-class life and secure retirement just as it did historically. Most Americans today have never stepped foot in a factory and have never heard about a vast range of trade types. Michael Nauman, the CEO of Milwaukee-based Brady Corporation, which as we'll introduce later in the book has transformed its factories into gleaming models of high-tech manufacturing, related recently that when his son told his friends, all in their twenties, about visiting him in a factory, they all asked, "What's a factory?" He lamented that a young local entrepreneur is much more likely to launch a coffee franchise than a machine shop, even though the latter has a far better chance of success.

The lack of understanding has made finding employees difficult for manufacturers. Brady noted, "There's a huge misperception . . . as to the benefits of industrial jobs. They offer stable employment, highly livable wages, and a career path. Yet we see so many two- and four-year college grads opting to become baristas instead of even considering a factory job." In fact, jobs in the Titanium Economy pay on average more than double the salary paid to workers in the service sector—$63,000 as opposed to $30,000 annually. Working conditions are generally comfortable, with well-lit facilities and top-of-the-line technologies, and many companies have innovative policies for workplace management, with many even being employee owned.

We believe that today the American Dream, that anyone can find success and economic security in this country through hard work, is as well-embodied by career opportunities in the Titanium Economy as it ever has been. A high school dropout can earn a GED while getting on-the-job training and then move right into a well-paying job with great benefits. A line manager without a college degree can earn enough to support a family of four while socking away a surplus for their child's college fund. A secretary working for a successful industrial company can retire with a $4 million pension fund. And if you're a hotshot engineer from MIT or you have an MBA from Stanford, while you can head to a big tech firm and work on an algorithm that may or may not change the world, there are barely one hundred such companies to choose from, and the competition for those jobs is fierce, to say the least. The Titanium Economy, on the other hand, has the highest number of open jobs at every organizational level across 4,500 industrial-tech companies that are the new linchpins of their communities. These are businesses where you can find endless, tangible ways to direct your talent and affect not just the lives of the people you work with, but of those around the country and the world—not to mention the life of the planet itself.

In short, opportunities for good jobs abound, with Titanium Economy companies typically generating more new jobs than can be filled. Much press has emphasized the threat of jobs being replaced by artificial intelligence and automation, but the need for skilled manufacturing workers is actually growing.

A Virtuous Cycle of Growth

The Titanium Economy is the secret weapon not only for workers and their families but also for improving socioeconomic conditions writ large. Supercharging this sector will send massive ripples throughout the economy, addressing many of the problems of economic growth and long-term poverty reduction that have stumped policy makers and elected officials for decades. The positive ripple effects are already seen in the scores of communities all around the United States that are home to Titanium Economy plants.

If you're wondering where to find these Titanium companies, look around you. Chances are good that one of the 4,500 is just down the road, or at most within a two-hour drive. For example, in addition to Brady Corporation, you will find that another 40 of them are scattered around the Milwaukee area. Any midsized city in the country has at least ten in the vicinity. Unlike the high-tech, financial, or even manufacturing industries of the past, there is no center of the Titanium Economy; it isn't fixed in any single geographic area. It's made up of industrials scattered across every part of the United States. And in each of these communities, Titanium Economy companies are creating a phenomenon, which we describe in the book as the Great Amplification Cycle: When people have good jobs that pay more than a livable wage, they can afford to buy a house, go on vacation or to the theater, and dine out in nice restaurants.

And as their communities in turn thrive, thus gaining a reputation as good places to live, different types of people will come to live there, diversifying the pool of talent. Those creative and ambitious types will in turn drive even greater innovation because with different perspectives come different ways to solve problems for customers. And that, of course, leads to a virtuous cycle: more profits and more wealth to be spread around.

In addition, because many of these companies have been in their communities for a long time, they're mindful of their place there, hiring from within families and generations, with fathers, mothers, sons, daughters, aunts, nieces, and nephews all working in the same company. Not surprisingly, with that longevity comes upward mobility, with many people having multiple careers inside these companies. We have met quite a few CEOs and senior management team members who started out on the shop floor—and we use the term not as a euphemism but as a badge of honor.

Companies that have formed the backbone of a town or a region for a generation or more, and who intend to stay there, have a vested interest in making sure that the health of their areas is not just served but bettered. Many are family owned, and those families take pride in supporting the health and overall welfare of their communities. One result of the strong connections these companies have made with their communities, as we'll explore more in subsequent chapters, is that some of them have driven important environmental innovations out of a strong sense of commitment to taking care of the places their workers call home.

Regarding the larger, national ripple effects, we believe the Titanium Economy has the potential not only to restore

America's might as a manufacturing leader but also to balance the economic scales in the United States, stemming and eventually reversing the growth of income inequality that has aggravated social divisions in recent decades. Studies highlight how directly intertwined the fate of the US economy and this sector are. When industrial companies struggle or fail, the negative multiplier effect on productivity can be severe. A 10 percent decrease in average manufacturing value per capita is associated with a 13 percent increase in income inequality. Supporting the expansion of the Titanium Economy will help remediate the growth of income inequality in recent decades, restoring the backbone of the US economy by broadening opportunities for wealth creation to those left behind. But that will only happen if enough people, from workers to investors, embrace its potential.

The Time Is Now

The United States has little time to lose. We estimate the next five years to be decisive both in protecting and investing in the advances made by the Titanium Economy. Many towns in our nation still have one foot in their economic graves while American innovation is creating wealth in other countries. Semiconductor manufacturing thrives in Taiwan while CATL, a Chinese company, controls the global electric vehicle battery market, producing batteries for many of Tesla's vehicles, as well as numerous other US brands.

America is at a crossroads in meeting the Titanium Economy's challenges, just as we were at the crossroads of the internet economy during the mid-1990s. The question is quite simple: Will the United States double down or fall behind?

Unfortunately, the continued misunderstanding about the nature of the industrial-tech sector is underscored by the undervaluation by financial players, and underappreciation of the innovation and potential for great careers in the sector is preventing us from fully capitalizing on the benefits. Consider that not even 1 percent of venture capital goes into industrial technology today. These are the problems we aim to rectify because there has never been a better time to seize the moment and strengthen our industrial base.

Once you know where to look, you will see the Titanium Economy and the opportunities it generates at work all around you. The time to act is now, and we believe the chapters that follow show the way. We'll introduce you to some of the key players in the Titanium Economy, show you how they create remarkable value through relentless innovation, and provide a playbook for catalyzing the extraordinary future growth that can make the United States once again the undisputed world leader in manufacturing, assuring that other countries don't grab the competitive advantage.

Chapter 2

The Home Court Advantage

At the launch of Tesla's Model Y in 2018, Elon Musk memorably said, "The issue is not about coming up with a car design—it's absolutely about the production system. You want to have a good product to build, but that's basically the easy part. The factory is the hard part."

Musk knew this from personal experience after working day and night alongside his team in makeshift tents to fulfill orders for the Model Y. As we write this, Tesla, along with other automakers, faces yet another crisis. Since 2021 an acute shortage of semiconductor chips has threatened to shut down a fragile economic recovery from the Covid-19 pandemic.

At first glance, you might think that the chip shortage would only really affect the computer industry and big tech companies like Apple and Microsoft, but chips control a host of essential products we take for granted. In cars, they regulate speedometers and antilock brakes. Your credit card now comes with a chip that changes the security protocol with every transaction to prevent fraud. You'll find everything from washing machines and smoke detectors to light bulbs and toothbrushes studded with chips. And without them, critical health devices, such as pacemakers, MRI devices, and blood sugar monitors, can't operate. Anything "smart"—the devices that rule so much of our lives in this day and age—relies on a chip.

In February 2022, the federal government warned that US manufacturers had a five-day supply of chips on hand compared to a forty-day supply in 2019, due to supply chain challenges. This is a problem that remains critical. Officials are working with port operations, unions and retailers, and industrials to unblock supply chain delays, but as a Gartner analyst noted, "Foundries are increasing wafer prices, and in turn, chip companies are increasing device prices." This means rising prices for manufacturers; that is, if they can get hold of chips in the first place. Even the best-case scenarios anticipate significant shortages for an extended period. Regarding cars, for which consumer demand outstrips supply, many customers are paying "supply chain fees" of a few thousand dollars for each new car. Compared to 2021, new car year-over-year inventory was down by 62 percent on US dealer lots in January 2022.

Even mighty Tesla announced that chip shortages meant it would not introduce any new products in 2022 and pickup truck enthusiasts awaiting its EV model would have to wait.

How, then, did Tesla crank out almost one million electric cars in 2021 during an unsettled period in which the supply chain was in disarray?

The company had a stockpile of chips, albeit ones that were incompatible with the cars they were currently producing. When they couldn't get the right ones, Tesla engineers zagged instead of zigged: They rewrote the software on their cars to work with the existing supply of chips on hand. Tesla could take this unexpected step because the company employs its own cadre of computer programmers. They went into the software to adapt it to alternate hardware. That's an example of Tesla's ongoing quest to take vertical integration, in Musk's words, to "absurd" heights.

More manufacturers are coming around to this way of thinking, opting to make their own parts where possible, as well as acquiring suppliers and bringing manufacturing home. Accelerating this trend, the supply chain crisis was a wake-up call for manufacturers who got caught empty-handed and had to reevaluate their supply chain. Indeed, entire industries are realizing that the cost savings from offshoring can sometimes mean an unacceptable loss of control.

We've certainly seen that scenario play out with semiconductor chips. Now, not only are manufacturers unhappy, but the US government has also begun to appreciate how dependent the economy, not to mention the military, has become on foreign suppliers. Intel, which ranks among the world's largest chip makers, is being urged to step up its production and advance US production capacity. Intel plans to build two new chip factories with a price tag of $20 billion to ensure we are not caught short-handed again. The good news about today's shortages is that they are fueling investments.

The renewed push to bring back the chip industry reflects what is happening in many sectors. Today it's computer chips, and tomorrow, with the right investment, it may be lithium-ion batteries, where China is eating our lunch, or the critical business of advanced photonics production, in which Europe has a near monopoly. Advanced photonics are critical for precision manufacturing and, indeed, the entire digital economy, enabling displays on cell phones, tablets, and computing devices.

As more advanced tech becomes prevalent in industrial products and processes, we predict more jobs returning to the United States. Why? Increasing product complexity and precision manufacturing require more oversight, and great gains in quality control are achieved when engineers and scientists are close to the factory floor. Many Titanium Economy companies have understood the value of keeping operations at home and have turned their factories into gleaming havens of high-tech manufacturing that offer workers, from the factory floor to the design and engineering teams, superior working conditions.

A New Breed of Factory

Walk through the doors of Brady Corporation's main production plant on Good Hope Road, on Milwaukee's northwest side, and the first thing you notice is the noise, or rather the lack thereof.

Down a hallway, you find the main room, where an array of boxy machines performs a variety of functions: They're punching, printing, stamping, and assembling behind noise-dampening plastic shields. A handful of employees dressed in the sort of polo shirts and chinos you'd see in a Gap commercial oversee the machines via LED monitors. There's no need to shout over the clanging din of a traditional assembly line.

Instead, the sound of human conversation and laughter echoes around the clean, well-lit factory floor. Soft LED lights shine on the walls in accents of soft blues and cream. Products are stacked in tidy piles you'd expect in a high-end outlet store. In fact, the place exudes the vibe of a Home Depot.

Brady Corporation is one on a long list of Titanium Economy companies that have transformed their factories into pristine, employee-friendly environments. Along the way, they have achieved both lean process improvements by incorporating advanced technologies into their workflow and energized employees by making the available jobs eminently more appealing. While media coverage of the manufacturing economy was focusing on the shift to China and the decline in the so-called Rust Belt, many American manufacturers were reinventing just about every aspect of their business.

Not only have factories become a great deal more efficient, but many Titanium Economy companies have implemented processes that allow their customers to interact with them in novel ways. They've also introduced methods of product design and manufacturing that involve more effective collaboration between designers and engineers and open up ideation about product innovation to the broad base of employees.

Brady Corporation CEO Michael Nauman has driven this transformation for facilities that once featured hulking contraptions that had to be repaired with wrenches and screwdrivers and were controlled by knobs, levers, and dials. Now, Brady's automated machines, sealed behind their soundproofed cases, feed operational details digitally to operator screens. "It's like your automobile," Nauman said. "When I was growing up, I could climb into my 1974 Chrysler Newport to work on the engine. Now your car motor is sealed with a label, warning, if you break the seal, the warranty is void. So, I can't even tell you what's under the plastic. It's the same thing with our new equipment."

Under Nauman's leadership, Brady Corporation, which was founded in 1914 as a producer of promotional calendars, has grown into a leading manufacturer of high-performance labels, safety devices, printing systems, and software for the electronics, telecommunications, manufacturing, electrical, construction, aerospace, and healthcare fields. With about 5,700 employees, the company's annual sales now slightly top $1 billion, and Brady has a market cap of about $3 billion. That success has been built by the kind of continual innovation and adaptation to market needs that we've seen distinguish so many Titanium Economy firms.

A big advance came during World War II when the company discovered an ingenious way to clearly differentiate electrical wires on ships, planes, and other equipment from one another, for safety in making repairs while also protecting wires from extreme conditions. Brady's patented self-adhesive, preprinted strips, which attached to wires, could withstand conditions from the 150°F tarmac heat to subzero temperatures at 37,000 feet.

Brady rode the success of this product to create a line of industrial identification products, which grew exponentially in the post-war boom—a trajectory that would continue over the next several decades. While the wire strips are still in service, such as for one of the company's major customers, the US Department of Defense, Brady's customer base now includes a wide range of industrial customers. If you've had a baby in a hospital recently, your child may have worn one of its identification bracelets, which assure access by only certain people to your newborn. Brady's identification labels embedded with holograms and other security features are also used by pharmaceutical companies to prevent the substitution of counterfeit drugs.

Both of these products, and many others, were innovated during Nauman's tenure, which began in 2014. Nauman also pushed to transform Brady's plant floors as workplaces and customer-facing environments that reflected a new outlook. Soon after he arrived, Brady teams went to work overhauling the factory, not only installing new equipment, but renovating decrepit, gloomy-looking buildings so they exuded positivity. "We painted walls that hadn't been painted since 1960," Nauman reports. He also had LED lighting installed, which is healthier for the eyes in addition to being more efficient. "If you've got a dark, dingy environment," Nauman said, "I don't care what technology you have, people don't perceive your facility that way. It's not only important that you have technologically driven products and processes; you have to project it. Prospective employees, our customers, and our suppliers come in and say, 'Wow, these guys are serious about this place being technology-driven.'"

Nauman has also used technology to offer customers enhanced service. Using a combination of robotics, automation, and AI, Brady offers a touch-and-click interface that allows customers to design their products online, such as electronic labels, locks, and custom signs, tweaking to their heart's desire and then obtaining a cost quote. Once the product is ordered, the instructions are automatically sent to the queue on the manufacturing floor. Little robots run around the factory picking up bins of products and delivering them to the loading dock for fast shipment to customers.

"You are going to see from us, and companies like ours, a lot more integration of software and customer interfaces," Nauman predicts, adding that customers increasingly want easier user interfaces and the ability to adapt their orders on the fly. Integration of systems is critical for offering this custom service and obtaining maximum efficiency. Nauman adds, "Our e-readers, scanners, optical scanners, and printer systems all operate together, so products can move through the manufacturing plant with minimal friction."

At the same time, customers still have the ability to talk to a representative any time they need to. "We get really high customer service marks because our customers can still reach people, even though we have chat bots and other AI responders," Nauman said. "Any company that doesn't offer a live person to problem solve with is making a big mistake."

Understanding that increasing automation will enhance many jobs, Nauman told us about an employee who maintains a machine that he helped create fifteen years ago. "No one else has ever touched it," Nauman said. "He has pride beyond belief." Nauman also expressed a belief in the kind of ground-up employee participation in innovation and customer relations that we've found is common in Titanium Economy companies, saying of this employee: "It's crucial that they are an integral part of the process of any change, as well as any customer service experience, and participate in any decision-making."

Highly Engaging Employees

At Graco Inc., every employee is encouraged to participate and be rewarded for finding ways to co-create the next best thing for their team or department. The company is a designer and manufacturer of fluid-handling products and solutions, pumping peanut butter into jars, gluing soles to shoes, spraying the finish on your vehicle, and more. Every quarter, it runs a contest that all employees can participate in, submitting suggestions for improvements in any aspect of the company that touches their work. Graco issues cash awards for the best suggestions and announces those winners to the whole company. Call it the village barn-raising method. The all-inclusive culture at Graco is a boon for employee engagement.

Graco has repeatedly landed on *Fortune*'s top ten list of Best Large Workplaces in Manufacturing and Production, which is based on the feedback of thousands of employees through the Great Places to Work survey. For Graco, the survey shows that nine out of ten employees say it's a great place to work. That's the kind of math that gets the company plenty of attention from job seekers.

We bet that few of you have “flow control” on your lists of hot sectors to watch, but this significant niche has outperformed other sectors for years. One player leads flow control: Graco.

In 1925, Russell Gray was managing a parking lot in downtown Minneapolis that also offered a greasing service. But that became challenging during the cold Minnesota winters, as the grease turned hard and nearly impossible to pump by hand. To solve this problem, Russell designed an air-operated, pressurized grease gun. He and his brother, Leil, turned the concept into the Gray Company, now Graco Inc., which was founded in 1926.

Flow control remains one of Graco’s core businesses today. It’s the technology that puts peanut butter in Jiffy jars, pumps ink into pens, and ensures the right amount of medicine goes into each gel cap. Numerous manufacturers need it. As a result of taking its role in the supply chain of so many other companies very seriously, the Minneapolis-based conglomerate has become a big fish in this very specialized pond, achieving returns that have outperformed those of Facebook and Google from 2015 to 2020.

That responsibility to provide precision performance looms large behind every decision President and CEO Mark Sheahan and his team make, and it was a key reason Graco didn’t follow so many other companies that moved production overseas. “You take a lot of the risk off the table too if you can control your manufacturing destiny,” Sheahan said.

Executive Vice President of Operations Angie Wordell, who heads all manufacturing operations as well as Graco’s oil and gas business, described the holistic control of Graco's operations. Wordell knows every inch of Graco's factories. Having worked her way up from an internship in 1996, she has spent time managing most of Graco's business segments. "We start with raw materials," she explains. "We do a significant chunk of our machining. We then move through material forming, processes such as paint and other coatings, welding, and finally assembly and distribution to our customers. All of our divisions operate in this same mold—this same approach."

While Graco runs one factory in Suzhou, China, a wholesale shift didn't make sense given that the company's product lines are broad and diverse, with many niches. "We are not making millions of anything," explained Sheahan. "Some manufacturers outsourced their products because they had so much volume that it made sense for them to attack it with labor, whereas we have a product line that's a mile wide and an inch deep." While the company could reduce labor costs with production in China, the wide array of products meant they would have incurred significantly higher shipping and logistics without any scale benefits.

In addition, the company culture is true to its DIY roots. Every product they make undergoes an extensive process of evaluation to determine whether it should be made in-house or outsourced. While it's more expensive than farming certain sections out to low-cost producers, along with ensuring that all parts meet the company's exacting standards, this system allows the company to customize when they need to. Even purchased parts get the Graco touch. Holding up a piece of metal that has been purchased from a competitor, Wordell explained,

"We will use a computerized, numerically controlled machine to process a piece like this to a higher quality all within the limited lead time required by our customers."

A key reason Graco employees are highly engaged is that the management fosters a culture of collaboration. One way it does this is by making sure that the company's engineers work in tandem with its designers. "Design never just throws something over the wall to engineering," Wordell said. "Instead, both groups work together on the premise that if design can think it, then engineering can make it, but that it can really only be accomplished as a team." This cooperative problem-solving between design and engineering is a core of Graco's success, a fluid relationship that is Graco's secret weapon. At so many manufacturing companies, by comparison, disconnects in communication and views about product development between design and engineering lead to costly delays in production, to say nothing of considerable interpersonal friction. At Graco, thanks to better teamwork, production concerns are embedded in the design from the start.

Collaboration in innovation is also fostered among workers on the factory floors. Each factory has its own process that Wordell relates. Typically, the team selects a challenge and then invites floor operators to come up with improvements. "Let's say we call for a safety challenge. Employees are asked to come up with ideas and how to implement them to improve safety. We also have quality challenges, cost-savings challenges.

"As a result, we always get suggestions that we can implement. Always. We then write up the challenge and share with the other factories where they learn and implement. This does two cool things. We challenge people to think outside the box, and it also embeds continuous improvement in the culture," Wordell said. That ethos stems from Graco's earliest days, when field salespeople would work with customers to solve problems and create made-to-order products. But that hands-on, on-the-ground problem-solving begat the modern day.

With each business segment encouraged to innovate, Graco's teams produce a constant stream of new applications and products. "We put together this grid that shows the product evolution by our business segments over the decades," Sheahan recounts. "It looks like a spider web. We did one product, and then this one evolved out of it, and then we took that technology and brought it to a new industry. The whole product evolution of the company has really been the core and what's driven most of our growth." This organic innovation is not only great for business results, it also fuels employee pride in working for the company.

Protect the Core

Another Titanium Economy player that has benefited from the wisdom of making its own vital components is Watts Water, which was founded in 1874 as a tiny New England machine shop and has grown into a multinational company that sells products and systems to manage the flow of water and energy in commercial and residential buildings. With approximately 4,000 employees, Watts posted $1.8 billion in annual revenues in 2021 with a market cap of $5 billion.

Watts Water owns a foundry and machine shop in Franklin, New Hampshire, which opened in 1959. In 2013 Watts expanded, opening a new 30,000-square-foot foundry that makes unleaded brass, an essential component of the company's products. When Robert Pagano joined as CEO in 2014, his company-wide review found that the foundry was not performing optimally, especially given all the new technological and process advances in the sector, and he therefore made the prescient decision to further upgrade what had already been a significant capital investment made in the foundry prior to his tenure. He solicited specific and relevant expertise from outside, including from his former company, to help stabilize and enhance the foundry. In addition, Watts spent a lot of capital to upgrade and automate its machine shops.

"The foundry is now one of the most important lead-free plumbing foundries in the US," Pagano told us, "because nobody else in the US is doing it." Originally the result of Pagano turning the telescope around to look at the business from the customers' perspective, the foundry and local machining capabilities had the added benefit of getting Watts through the current supply chain crunch. "When Covid came along, with all the supply chain disruption, we leaned on our foundry and machine shops in North America, which allowed us to keep up production during a truly challenging time," Pagano recalled. "And by focusing our teams on serving our customers, we were able to get them the essential products they needed."

Pagano's leadership represents how limber the transformation of operations within Titanium Economy companies can be. It's also a model for those firms not yet availing themselves of the benefits of the vertical integration and strategic focus that are driving such growth for others.

When Pagano arrived at the company, he told his team, "You have to protect everything around the core first. Only then can you expand." Managers were hunkered down in a defensive posture relying on legacy operations and siloed sourcing practices. "When I started," Pagano explained, "we had thirteen new product development processes, and no one was following them. And they thought the voice of the customer was the reps, not the real end-user customer. So we had to change that."

Part of the strategy for remaking Watts was to streamline and standardize the company's processes and to consolidate, where possible, existing Enterprise Resource Planning systems. In addition, the company added a business intelligence system that oversaw all the existing ERPs to allow Pagano's team critical access to information about product, customer, and channel profitability. Pagano said, "This was critical. Now we have standardized reporting and data and we can train more people to talk the same language and have the same standards, versus everybody doing their own thing."

Pagano also installed a new management team of executives who embraced innovation, modernization, production resilience, customer service, and operational excellence. "In the year before I arrived, only the sales team received bonuses, and management received none," he told us. "Objectives and incentives weren't aligned." Additionally, Pagano instilled a disciplined approach to acquisitions. "We were milking the core business to do the next acquisition. We were number one, two, or three in everything we did. Yet everybody was just pecking away at us because we were distracted by all the acquisitions versus focusing on modernizing our core products." Recalling that it took up to a year for a $10 million capital request to get approved at his old company, Pagano aimed to speed up decision-making at Watts. "Now we can do it in a few days," he said. "My point is: you've got to be nimble, you've got to be fast, you've got to make decisions," even when you don't have "100 percent of the information. But if you can get at least 85 percent and have an experienced team behind you, things can move quickly. If you make a mistake, just quickly admit it, be very transparent, and then fix it."

Watts Water continues to thrive, as its performance over the past five years has shown. In fact, its total shareholder returns from 2015 to 2020 grew at a 21 percent CAGR, which puts its performance within the ranks of FAANG.

Although Tesla has grabbed the media spotlight and Wall Street's attention for its visionary model of production, the company is not alone. The agile leadership at Watts, the perseverance of Brady Corporation, the can-do spirit at Graco, and the embrace of innovation to drive value in so many Titanium Economy companies give us confidence that the United States really can deliver the promise of a new age of industrialization and become home to 1,000 companies like Tesla.

Chapter 3

The Long Haul

At dawn on a chilly April morning in 1975, Doug Casella climbed into his pickup truck, cradling his cup of Dunkin' Donuts coffee, turned the ignition, and began making his rounds picking up garbage from a handful of customers in Rutland, Vermont. He bought the truck with money saved from odd jobs he'd done in high school and the proceeds from selling his car. The purchase turned out to be a wise investment, as Doug's fledgling enterprise quickly took off.

Before long, Doug told his older brother, John, "You should come be my partner," and John, who was trying to make his way in real estate, sensed a good opportunity. Also lured by the easy camaraderie the pair shared, he didn't hesitate. As with so many Titanium Economy founders, the brothers have remained firmly in charge after taking the company public in

1997, with John now the CEO and Doug continuing to steer the company's future as a member of the board.

Trucks with the Casella Waste Systems logo emblazoned on their side are now a common sight in the Northeast, where the company is not only the premier waste disposal firm but also a leader in environmental, social, and governance criteria (ESG) with innovation in recycling and regenerative energy production, harnessing the potential of breakthrough technologies—like AI-enabled machine vision—to dramatically advance the goal of a fully sustainable and waste-free future.

The Casella brothers have succeeded due to the relentless pursuit of innovation sweet spots, discovering golden opportunities that larger firms in their sector mostly overlooked and combining technological sophistication with good old-fashioned engineering stick-to-itiveness. A sustainable future depends on not only dealing with our waste in safe and environmentally friendly ways but also finding ways to repurpose waste, including turning it into clean energy and fertilizer, as the Casellas are doing, and into many other types of new products.

Yet the ingenious contributions of countless innovators, and the impressive revenues and returns they've generated, have been largely overlooked by the media. While advances in waste management don't make the nightly news, and recycling is often characterized as economically impractical—and even regularly reported to be "dead" in headlines—Titanium Economy innovators like the Casellas are proving the waste business is, in fact, ripe terrain for enviable profit-making, as well as for planet-saving work.

The Casellas' journey from their two-person operation to Titanium Economy standout didn't require chasing venture capital, or generating lots of press hype, as featured so prominently in many Silicon Valley origin stories. It required the dedication of learning every inch of their business and building strong relationships with customers and other Titanium Economy innovators. They then leveraged those resources to continually push for breakthroughs. The business was built, in short, on hard work, which never intimidated the Casellas.

"We grew up mixing cement, carrying bricks," John recalls, adding that the brothers' father "was a taskmaster, to say the least." They helped their dad build the red brick motel the family ran, and lived in, which was situated on Route 4 between rural Rutland and the popular Killington, Vermont, ski resorts. Their father, Raymond, and mother, Thurley, moved to Vermont from Yonkers, New York, in the early 1950s after having spotted an opportunity in the post-war ski resort boom that brought 55 new resorts to Vermont by 1948. Ray, who was a certified operating engineer, strung up lights so they could work at night, and the family lived in a tent on the property.

Running the motel was also a family affair, with John and Doug bussing tables in the on-site restaurant and shoveling the heaps of snow that would cover the vast parking lot in the winter. Ray and the boys also did construction work and found other odd jobs to help make ends meet.

Not surprisingly, Ray's uncompromising work ethic rubbed off on his sons. In short order, the brothers led the hauling business in central Vermont. They spent their days picking up trash and their nights doing the business's paperwork, and it wasn't until the number of routes they serviced expanded beyond their wildest dreams that they left the driving to others. Being so hands-on helped them to develop strong relationships with the communities they served, and they were well-known around Vermont as the "odd couple" due to their contrasting personalities. Doug was hot-tempered and wore his emotions on his sleeve while John was more diplomatic. They were unified, though, in their vision for growing the company—quick to adapt to emerging opportunities and nimbly riding the rollercoaster of economic turmoil in the late-1970s and 1980s.

Their early success was met with inevitable resentment from some of their competitors, who spread rumors that the company owed allegiance to the Mafia, which was known to be involved in the business in some cities. When that didn't stick, the new story became that they were controlled by nameless, faceless corporate taskmasters. The truth of their success was that they immediately began innovating.

In the years when they started, the waste management industry badly needed an overhaul. The average hauler couldn't see beyond the horizon of the local garbage route and landfill dump to the bigger strategic potential of recycling, hazardous materials cleanup, waste reduction, and regenerative energy production. The Casellas, by contrast, appreciated that the social landscape was changing. Driven by qualms about pollution, unregulated landfills, and unsafe water, consumers were increasingly interested in the three Rs: reduce, reuse, and recycle. Just two years after the company's founding, the brothers purchased a baling machine from an old mill in Maine for $3,000. They used it to compress cardboard boxes into large bales for shipping to recycling facilities, thereby opening Vermont's first recycling center.

During the 1990s, the brothers poured their entrepreneurial energy into acquisitions, purchasing fifty companies, amassing routes that included over 68,000 customers in the process, and establishing a network of four landfills and eight recycling facilities. Along the way, the Casellas fell into the trap of overdiversifying the business, as well as taking on excessive debt. The company initially funded the growth via a series of private investors before leaning on stock sales after it went public in 1997. Some of the companies purchased fell outside John and Doug's areas of expertise. New Jersey–based KTI, a waste processor with operations in other industries including insulation production and tire recycling, was one such company, and the brothers soon found themselves overextended and under fire. The stock price plummeted 70 percent in the year after the KTI acquisition, and despite attempts to right the ship in the early 2000s, the 2008 financial crisis delivered another major blow as demand for waste collection and disposal suddenly decreased.

In the face of those challenges, the Casella brothers realized that the way back to financial success was to focus more tightly again on the core business of waste collection, disposal, and recycling that they knew so well. The company shed noncore operations. John, by then the company's CEO, also adopted management innovations, revamping the executive team and decentralizing decision-making, giving managers more flexibility to address issues as they arose. When the company faced a takeover attempt, the Casellas were ready, beating it back in large part thanks to the restructuring measures they'd undertaken. The strong support they had built within the communities they serviced, including being avid champions of several charities, contributed to shareholder confidence in their leadership.

Soon, the stock price began to surge again.

In the years since, the Casellas have increasingly leaned on growing their green technology operations. They've built the single recycling center they opened in 1977 into a network of twenty-five facilities across the Northeast, and they've been at the forefront of innovation in the business. They were quick to work with Recycle Bank, a firm that created a program allowing customers to track how much they recycle and earn coupons according to that volume at stores like Home Depot or Starbucks. The offering was a huge success in their communities. But the Casellas knew it would take more than such incentives to move the recycling needle more forcefully. They were also quick to appreciate the advantage of single-stream recycling, which greatly increases compliance with recycling initiatives.

As old hands at the recycling game, the Casellas knew all too well that many people don't recycle because they are put off by all the necessary sorting. When they learned that some recycling operations were switching to a system in which customers simply put all their recyclables in one bin before they were later sorted at the collection facility, the Casellas created their "Zero-Sort Recycling" program, which allows customers to "fill the bin and leave the rest to us." Through such foresight, the company has recovered over 800,000 tons of recyclable materials and counting. But the Casellas aren't resting on their laurels.

They've kept investing in the best new technologies for making recycling more efficient and impactful. The Casellas' entrepreneurial spirit, along with their work ethic, enabled them to capitalize on shifting tides in the industrial space even as scores of legacy companies struggled to adapt to the changing landscape.

Automating the Dirty Work

The Casellas have recently invested in state-of-the-art robotics machinery similar to that which is manufactured by fellow Titanium Economy maverick Bulk Handling Systems (BHS), based in Eugene, Oregon. Founded in 1976, BHS makes the work of sorting recyclables both more accurate and more humane.

We've all been guilty of looking at a piece of trash and not knowing if it's recyclable before tossing it in the recycling bin, half in hope and half in frustration. And precisely because we've all done it, an untenable problem has arisen. These improper items muck up the entire recycling system. Grease from pizza boxes jams gears; garden hoses become snakes coiled around equipment; unwashed peanut butter jars soil surrounding items, causing entire loads to be rejected; and flammable gas canisters cause fires and put workers at risk of injury. Workers at most recycling centers pick through the morass, separating recyclables from unrecyclables, which Steve Miller, CEO of BHS, considers "the worst job in the world, the last job anyone wants." What's more, offending items sneak past workers because they can be hard to spot. The result is that machinery often needs to be shut down for repair, cutting significantly into productivity.

In response, BHS created one of the industry's most advanced automated recycling systems, which relies on artificial intelligence to sort and identify recyclables. The company makes optical sorting units that use spectrometers to "see" items by evaluating what they're made of, with specific materials having distinctive spectrographic "signatures." Meanwhile, the company's Max-AI robotic sorter uses a multilayered neural network, the building block of advanced artificial intelligence capabilities, to recognize objects based on extensive training, called machine learning. A robotic arm then sorts items into bins by category of material and extracts unwanted items, reducing the need for people to stand on sorting lines doing the grunge work. Employees instead primarily do the less-dangerous and -dirty work of transporting bundles of recyclables around facilities and monitoring the equipment. The machines pay for themselves in two years.

Not surprisingly, BHS's business has grown rapidly since the machines went on sale in April 2017, with more than 200 neural networks in use in the United States and around the world. Moreover, since 2005 under Miller's watch, the company has grown from 30 to 300 employees and generates north of $100 million in revenue each year.

The Casellas have also invested in remarkable machinery for cutting down on the amount of organic matter sent to landfills, which makes up almost a quarter of all waste buried in them.

Much of that is food waste, a good percentage of which is still in its original packaging, never opened, such as cans of soup or boxes of crackers that have often been tossed because they've passed their expiration dates. Up until recently, these items couldn't be recycled because separating the contents from the packaging would have been so time-consuming that it wasn't economical. The volume of this waste is staggering. In Vermont alone, which is the second-least populous US state after Wyoming, it amounts to about 80,000 tons a year.

In early 2021, the Casellas unveiled the state's first depackaging facility. It features a giant red machine called a Turbo Separator, nicknamed Thor, which is manufactured by Scott Equipment Company, based just south of Minneapolis, Minnesota—another family-run Titanium Economy leader. The machine features paddles that break open the food packages, freeing the food matter. Robotic arms then pluck out the broken packaging for recycling, or the landfill, depending on the material. Meanwhile, screens filter out the food waste for composting.

Thor assuredly handles everything from plastic-wrapped stale bread to off-spec Keurig coffee pods and Ben & Jerry's ice cream containers. We could hear the satisfaction of saving so much valuable material from going to waste in the voice of John's son Michael, who is Casella Waste Systems' general manager, when he explained to us that "once that product can't be used for human consumption, we can actually take it instead of sending it to the landfill and process it and then turn it into a recycled product." That might be new packaging, as well as the compost from the food matter. The Casellas are also engaged in a potentially transformative innovation for turning another type of organic matter into energy.

THE HOLY GRAIL

Since their first foray into recycling in 1977, the brothers had pondered the holy grail of waste management, wondering, *what if waste could be turned into energy?* Clean energy, that is.

They're finally realizing that dream, and it's happening in a place that coincidentally bears the same name as their hometown of Rutland, though this Rutland is 120 miles away in Massachusetts and is home to a farm run by two equally close-knit brothers.

Jordan Dairy Farm sits on a generous swath of pasture that 800 cows call home. Not to be indelicate, but cows mean manure—in this case, 10,000 gallons of it a day. And that's where the twin silos that stand guard over the grazing bovines come in. Rather than storing grain, these silos are anaerobic digesters, producing electricity from the copious volumes of cow manure the herd produces, as well as food waste collected from far and wide. The Casellas have sponsored this pilot program on the farm in partnership with Vanguard Renewables, based in Wellesley, Massachusetts, which specializes in working with dairy farmers to adopt renewable technologies like regenerative agriculture and anaerobic digestion.

A column inside the first silo receives forty-five tons of food waste from a dozen organizations—including discarded hotdogs from nearby Fenway Park and meatballs, pizza, salads, and food scraps from local restaurants, manufacturers, and schools—along with twenty-five tons of cow manure generated by both the Jordans' cows and those of other nearby dairy farms. Instead of the methane released from the manure contributing to greenhouse gas emissions, it is combined with gases from the food waste in the airless silos and then funneled to a giant generator, which turns the witch's brew into electricity.

That energy is sufficient to power the entire farm plus 300 homes. Meanwhile, the leftover gunk in the digester serves as a kind of super-charged fertilizer, going back into the land to help grow corn and hay. According to Jordan Dairy Farms, the process has improved crop growth by 50 percent.

Keeping abreast of exciting new solutions as they have emerged has kept Casella Waste Systems nimble and vibrant and has provided high-quality jobs for a workforce that's grown to 2,300 employees across Vermont, New Hampshire, New York, Massachusetts, Maine, and Pennsylvania while generating revenues of $800 million in 2020 with margins of 20 percent. Over the past five years, Casella has outperformed all FAANG companies with a growth of 60 percent CAGR in total shareholder return.

The Jordan Dairy Farm operation is one of a host of partnerships among agricultural enterprises, restaurants, supermarkets, businesses, schools, and municipalities working with waste management innovators to both power their own facilities and supply clean energy to the grid. That symbiotic problem-solving has been a core feature, not only of the

Casellas' approach to growing their business, but also of the Titanium Economy as a whole. Ecosystems of companies and customers support constant experimentation with emerging technologies, applying them to mutually beneficial advances.

As in nature's ecosystems, companies occupy specialized niches, honing their operations to constantly adapt to evolving needs that they have the depth of experience to identify and the creativity and drive to discover inventive solutions for. They vary greatly in size, with some dominating niches that, while specialized, are very large scale and growing. Many are helping to save the delicate natural ecosystems their symbiosis mimics.

Getting Down in the Dirt

"As any chemist knows, you can't destroy or create matter," said Scot Shoemaker, the vice president for engineering and operations for Clean Harbors. "But you can change its form into something less toxic."

A new Clean Harbors facility in El Dorado, Arkansas, is the first new commercial hazardous waste incinerator built in the United States to become operational in two decades. The $120 million facility, with the futuristic gleam of a space station, burns and processes industrial and laboratory chemicals, manufacturing by-products, medical waste, and other solid and liquid materials into a safe, odorless exhaust.

Vertical integration focused on hazardous waste disposal and recycling has propelled Clean Harbors' forty years of growth since its founding in 1980 from a simple office trailer to 750 locations. By 2020 the company was generating revenue of $3.14

billion while employing 14,000, making it the nation's largest environmental and industrial services provider. CEO and founder Alan McKim called his company the "Ghostbusters of America, doing jobs no one else wants to do."

McKim's career of taking on tough work that gets little press but addresses vital needs is exemplary of the zeal that Titanium Economy innovators have for deeply understanding the details of operations. The majority of industrial-tech players are legacy companies that have persevered through hard times and struck as opportunities arose. Spotting those opportunities is enabled by leaders who aren't afraid of getting down into the mud of problems, which in Alan McKim's case recently meant getting down into the tar and sludge of an oil tanker that needed to be cleaned.

One of the companies Clean Harbors has acquired in its vertical integration strategy is Safety-Kleen, the innovator of a brilliant solution for re-refining waste oil by collecting used oil from all sorts of vendors in tankers, refining it in their facilities, and shipping the clean oil back out in the same tankers once their insides have been made pristine again. McKim had gamely volunteered to try his hand at the grueling job.

"Are you afraid to get dirty?" the Clean Harbors supervisor asked McKim, not knowing that her new charge, "Bill Anderson," was the company CEO and that what the documentary crew filming them was supposedly making was really an episode of *Undercover Boss.* Posing as a former mechanic looking to restart his career, McKim tried his hand at multiple roles within Clean Harbors' facilities and job sites. Stints as a Class A truck driver, an industrial technician working in highly confined enclosures, and a field service worker responding to a hurricane-related disaster, McKim said, "allowed me to revisit my roots and see the company from the front lines again."

Born and raised in Braintree, Massachusetts, on Boston's South Shore, McKim got his start in the oil spill cleanup business after being hired by a local retired NFL player, Bob Dee, to work with him at Jet Line Services when he was a student at Northeastern University.

"Bob was like a dad to me," McKim said. "I started Clean Harbors with four people and a truck in 1980, a year after Bob died."

One of the company's first major jobs was on Cape Cod in 1984, when Clean Harbors pumped more than 100,000 gallons of oil off the crippled tanker *Eldia*, helping to avert an oil spill when the ship grounded off the coast during a snowstorm. Fast forward thirty years to what many experts consider the worst environmental catastrophe in American history: the April 2010 explosion of the Deepwater Horizon, the Transocean deep-sea oil rig in the Gulf of Mexico. That accident killed eleven people and released 134 million gallons of oil—the largest offshore spill ever in US waters. Scientists are still studying its environmental effects, which include the death of thousands of marine mammals and sea turtles and the contamination of their habitats along with 1,300 miles of coast.

Considering the global scrutiny and outrage following the accident, the US Coast Guard and a host of senior federal and state officials turned to Clean Harbors to undertake the large-scale response to an unprecedented crisis: an offshore spill zone affecting communities across three states. McKim and his Clean Harbors team had earned their credentials in the response to the Exxon Valdez oil spill in 1989, the cleanup at the World Trade Center site immediately after the 9/11 attacks, and the decontamination of NBC's offices when it was targeted with anthrax the same year.

"Those beaches are the prettiest you'll ever see, with the finest sand you could imagine," McKim said. "The idea is to try to fight the battle on the water and collect the oil before it comes ashore, and if it does come ashore, to try to mitigate the damage."

In leading the historically scaled Deepwater Horizon spill cleanup and environmental recovery effort and mitigation work, Clean Harbors also provided logistical support to the Coast Guard and health and safety training to residents recruited to assist in the cleanup along the coastline. Not only did the job require technical expertise, it was a massive exercise in people logistics.

"The best way I can characterize it," McKim recalls, "is that everything that goes into fighting an oil spill is no different from fighting a battle because we had 25,000 people taking that thing on. Its magnitude was enormous. There were command centers, and in those command centers you'd have to feed the troops and provide safety training, transportation, and housing."

As employees and volunteers decked out in protective gear wilted in the high heat and humidity while shoveling sand into containment vessels, Clean Harbors set up a hospitality tent at the central worksite, installed air conditioning, one hundred portable toilets, stacked up coolers with tons of cold drinks on ice, and provided breakfast, lunch, and dinner. All the while, McKim oversaw the entire operation, greeting everyone as they took breaks.

It wasn't until McKim went on *Undercover Boss*, though, that he truly understood the unseen hazards of the work, like encountering snakes while cleaning up in the wake of a hurricane in Texas or lacking a place to shower after climbing out of a dirty tanker in a re-refinery in Indiana.

To ensure his employees wouldn't recognize him, McKim received a makeover from the show's producers. Out went his trim haircut and corporate dress, replaced by a wiry ponytail, a mustache, and the hangdog demeanor of someone who'd been chewed up by life's hardships. "We did a test run with the makeup at my house with my kids and grandkids there," he recalls with a chuckle. "One of my younger grandchildren didn't recognize me and was afraid of the stranger in the house. I hardly recognized myself."

He saw firsthand the grit required to complete a full day's work, day after long day. As he cleaned the sludge from that oil tanker, he was covered in sweat and his hands shook with stress and exhaustion.

The Safety-Kleen acquisition, finalized in 2012, was the biggest deal in Clean Harbors' history at a cost of $1.25 billion. Safety-Kleen is the largest re-refiner of waste oil in North America, and the acquisition signaled McKim's seriousness in pursuing an underserved market: recycling motor and industry-grade oil, a nonrenewable source of energy that otherwise would be discarded in environmentally harmful ways. Just as in cleaning up oil spills, the work involves logistics comparable to those of going into battle.

The key to victory in the oil re-refining business is transportation logistics. Collecting oil from 2,000 auto service centers, fast lubes, car dealerships, factories, and other sites across the United States requires an extensive transportation plan. For this reason, Clean Harbors operates one of the largest private fleets in the country, and on any given day the company's trucks, tankers, and barges are busy picking up some of the 200 million gallons of oil per year they collect from customers and transporting it to its processing centers.

Breaking the "traveling salesman" code has long been the holy grail of route optimization. Route optimization requires factoring in millions of roadways, weather conditions, and fluctuating fuel costs, not to mention customer requirements as variables. The challenges of creating the best possible itinerary, with multiple stops before returning to the starting point, require a dizzying amount of data that stymies normal computers. With quantum algorithms, the company balances its far-flung internal fleet, as well as external rail schedules, to manage routes without compromising timely service to customers. Through such route optimization, tankers, which would previously have hit the road with half-full loads, can operate at near 100 percent capacity, saving on fuel and time. This is also why the tanks that McKim cleaned on *Undercover Boss* needed to be spotless. For greater efficiency, the same tanks that bring in dirty oil to be re-refined are quickly turned around, filled with sparkling-clean oil for delivery.

Route optimization has led to a 12 percent reduction in miles driven while transporting ever-increasing loads. These savings add up, and the acquisition has proved a brilliant move. In 2020, Safety-Kleen accounted for 35 percent of Clean Harbors' total sales.

Alchemizing Gold from Trash

As the stories of Casella Waste Systems and Clean Harbors exemplify, waste management is not for the faint-hearted; nevertheless, the problems posed by waste are among the most urgent we face, from the poisoning of all manner of species by toxic chemicals leached into waterways, to the clogging of our oceans with plastic refuse, to the release of greenhouse gases from cow manure, which accounts for an astonishing amount of annual methane emissions. They affect every single person on the planet. Titanium Economy companies are doing the bold work of creating and deploying solutions, making transformative on-the-ground progress while flying under the radar of the tech hype machine.

In addition, the stories underscore another aspect of the Titanium Economy. They're resilient. The narrative about industrials—that they have been on the decline for decades—comes partly from the fortunes of the auto industry that was generalized to the broader industrial landscape. Meanwhile, industrial companies not only survived but thrived and are poised to flourish in the technological future. We chose to focus on waste management because most people perceive it to be backward and stodgy when it is anything but that. If you had invested $1,000 in Clean Harbors at the beginning of the year

2000, your investment would have grown to $124,887 at the end of 2020.

And yet, we have only scratched the surface of this rich terrain. New technologies will continue to emerge, and new combinations of technologies will offer untold opportunities. Ambitious innovators could hardly do better than to set their sights on the pressing challenges of assuring we have clean air to breathe, clean water to drink, and enough healthy soil to grow the plentiful food that an ever-increasing world population will require.

Chapter 4

The Titanium Playbook

In the middle of December 1998, Larry Mendelson left his Miami home to board a predawn flight to LAX in Los Angeles. After touching down, he rented a car for the last leg of his journey, driving 50 miles southwest of LA to the city of Corona, population 157,000, the home of Fender Guitars and Monster Drinks as well as the innovative industrial-tech firm he had come to visit, Thermal Structures. The gregarious, almost always smiling businessman was uncharacteristically frowning. He'd just learned that at least a dozen other players were courting the company he'd traveled 2,700 miles to see. Not only that, but the word was also that their offers would go far beyond what he planned to pay.

By this time, Larry Mendelson had made dozens of acquisitions for HEICO, the company he had taken control of along with his two sons, Victor and Eric, in 1990. He knew that he couldn't win if it came to a bidding war for Thermal Structures because, unlike most other bidders, the money he planned to put on the table was coming directly from HEICO's coffers. While eschewing debt left him at a disadvantage when it came to the purchase price he could offer, it also meant he wouldn't feel pressure to slim down the operation for a quick profit, which might buy him extra points with the seller.

Mergers and acquisitions do not have to be a financially burdensome game or a blood sport. While Larry Mendelson was well aware that "show me the money" was the driving mantra for most sellers, he hoped the founder of Thermal, Vaughn Barnes, would prove to be an exception. Thermal fit the HEICO model for acquisitions beautifully, at least from what he'd seen so far on paper.

The strategy for growing HEICO that Mendelson and his sons have rigorously implemented demonstrates the playbook for growth in the Titanium Economy, which includes strategic acquisitions to lead a niche. HEICO had carved out a lucrative market making products for the airline industry that were too hard—or too expensive—for others to produce. Regulation and certification processes, along with expensive labor costs for qualified staff and high liability, make it difficult for new players to enter the market. You had to know what you were doing, and Thermal was one of the standout firms in the terrain. Among Thermal's products were metal shields designed to withstand the intense airflow of thrust turbine engines, much like the kind that powered the commercial jetliner Mendelson had flown on that morning. At full power, those engines could heat up internally to 590°F. Conversely, when flying at 39,000 feet, they had to withstand external temperatures that might dip to –137°F, with tailwinds of up to 800 miles per hour. Built to operate under such extreme conditions, shields that can take that sort of stress and keep an airline engine humming require exquisitely precise production.

Also appealing was how strategically Barnes had grown the company, branching out into specialty niches and making everything from fire-retardant insulation for jet aircraft and military vehicles to fire barrier systems for high-rise buildings in earthquake-prone areas. If you needed composite acoustic liners to dampen the noise of a jet engine or large industrial fans in your factory, Thermal Structures could custom manufacture them for you. That custom work requires in-house innovation, especially since many of the products the company makes require tools that can't be found on the open market. By necessity, Thermal had therefore invented such tools—a remarkable 35,000 of them in total. That talent pool of high-caliber engineers the company had attracted was most impressive. The company clearly had the means to stay on top of whatever engineering it would need, including keeping up with the innovation of its production lines. Machines can become obsolete fast, Mendelson knew, especially nowadays, and the quality of expertise for re-engineering those lines fit heavily into HEICO's calculation as to whether a company might be a good investment. Thermal also had a strong balance sheet.

The company was doing so well that Barnes, then only forty-five years old, decided he would be one of those entrepreneurs who took early retirement and traveled the world, so he put the company up for sale. When Mendelson learned

Thermal was on offer, he decided to pay a visit not only to kick the tires but to impress upon the founder and owner that he had something more valuable than the dear sums that others would no doubt offer.

Mendelson began forming a good impression of Barnes right away. Barnes greeted him when he walked into the lobby rather than sending an assistant. *Interesting*, Mendelson thought to himself. Barnes also wanted to take Mendelson on an immediate tour of Thermal's Skunk Works—not wasting any time.

"No two planes are exactly alike," Barnes told Mendelson as they headed to the shop floor, "even if they're both an Airbus A320 made at the same factory. Each shield needs to fit like a glove, and we need to constantly adapt them to accommodate flights that keep getting faster, some even hypersonic." Mendelson knew that the faster a plane flies, the more heat the engines generate. With hypersonic meaning five times the speed of sound, the shields needed to perform at a high level, withstanding the intense air flow of thrust turbine engines under extreme conditions. Shields that can take this stress and keep an airliner engine safe aren't stamped out in the machine shop down the street.

The two men quickly clicked. Both were numbers people at heart, Mendelson a one-time accountant and Barnes a trained engineer. But as they toured the Thermal facilities, Mendelson could see that they also shared a commitment to good people management.

To show Mendelson what his factory could do, Barnes guided him to a station where several programmed, super-hot lasers were busy carving out shells, and then to another station where noisy waterjet machines smoothed out any imperfections. "If it isn't exactly right, we start over," Barnes said. "We don't hesitate to scrap anything that isn't perfect."

Barnes introduced Mendelson to one of the line supervisors, and Mendelson asked her about her background. "When I started, I didn't know a punch from a die," she told him. She had started as an assistant to the safety technician before taking computer programming courses at the local community college at Barnes's urging. As a result, she ended up with a business degree, working her way up in the company after having "done a turn at pretty much every segment of the manufacturing line along the way." Mendelson was keenly attentive, listening to what she was telling him while also noting the enthusiasm she was expressing. "I feel honored and humbled to have this responsibility," she continued. "We know lives are at stake. Our shields can't fail."

Unbeknownst to Barnes, the encounter with the line supervisor checked an important item on Mendelson's list for evaluating possible acquisitions. As dazzled as he might have been by the fancy technology they were using, he was always far more interested in learning if employees were motivated and respected by the management for their skills and opinions. That Barnes had given her the opportunity to speak with him was a good sign. That he had encouraged her to develop her talents was another important indicator. Fostering the skills of employees, and providing long-term opportunities for them, were additional hallmarks of companies Mendelson sought to acquire.

The most crucial factor of all to Mendelson in evaluating Thermal was how well the company would fit with HEICO's culture. From the start, Mendelson had taken great care in building not only HEICO's portfolio of products but its organization. When he was young, he'd told his father, "I want to be rich," and the elder Mendelson had scoffed. "Don't think of how much money you're going to make. Find something you love to do. Then go all in and do it the best way you know how." Mendelson never forgot what his father had said and built HEICO on that simple ethos. He perceived that Barnes had also built Thermal with that kind of passion, and what's more, he had infused the organization with his pride and enthusiasm for their work.

As he and Barnes continued their tour, they began to talk money, and Mendelson pondered whether Barnes was the kind of guy who would take the highest offer and run. As if on cue, Barnes volunteered that Thermal had multiple interested parties. But Mendelson thought he just might have an ace up his sleeve in bidding for the firm.

The next day, back at his modest office, Larry Mendelson met with the six members of his executive management team to discuss their bid for Thermal. His two sons are on the team, having helped their father build the modern-day incarnation of HEICO and after working with him to gain control of the company. They had been instrumental in even deciding to look for a firm to run.

Mendelson had never set out to be an aerospace mogul, or even the CEO of anything, for that matter. The native of New York City had stayed near home for college, enrolling at Columbia University and earning an MBA. After a stint as a number cruncher at Arthur Anderson, once one of the country's largest accounting firms, he had made his way to South Florida and began investing in booming real estate, setting the stage for his next act. Condo conversions became his specialty. But after his sons followed in his footsteps in the 1980s by graduating from Columbia, where they studied business and economics, the three decided to work together.

"I always wanted to work with my sons," Mendelson said. "My father and I were close. He'd retired by the time I got out of college, and I never got to work with him. I wanted to change that with my sons."

They went hunting for a company to buy, and in 1989 they found what they saw as their best target yet in HEICO. Larry Mendelson wanted a company that made something but also had a solid balance sheet. HEICO had begun operations in 1957 and went public in 1960, but it had grown only sporadically since then, which the Mendelsons decided was the result of "unenthusiastic management."

"Board members owned very few company shares," Larry recalled. "They had no skin in the game." Still, HEICO had a solid foothold in a niche of the aerospace market as a result of its 1974 acquisition of Jet Avion Corporation, which supplied FAA-approved parts for aircraft engines. This was a sector that the Mendelsons, after much study, decided had great upside potential.

They bought $3 million in HEICO shares, using half cash and half debt, and with their stake demanded a seat on the HEICO board. After being rebuffed and launching a protracted proxy fight, which ended up with the court deciding in their favor, the Mendelsons won control of the company in

January 1990. They swept out the old management and reconstituted the board, putting themselves in charge.

"Thermal Structures sounds a lot like us many years ago," Mendelson told his executive team. He also told them he especially liked Thermal's growth track record and history of rich cash flow. Beyond that, the team agreed that keeping Vaughn Barnes was crucial and settled on offering him a 20 percent stake in the company that he'd done such a stellar job building and, crucially, running day-to-day. Mendelson ended the meeting feeling bullish. Now he just needed to convince Vaughn Barnes the offer was a good deal and that he should stay on to keep running the firm.

WHAT HAPPENS TO MY PEOPLE?

Vaughn Barnes was in engineering school when, like Bill Gates and Steve Jobs, he had an epiphany that led him to drop out. School could wait, but a burning idea couldn't. While waiting for a flight to take off from his native Southern California for a vacation in Hawaii, Barnes wondered how the metal cylinder that housed him and the other passengers kept cool at their flying speed of 500 miles per hour. That question occupied his mind during the entire four-hour flight, and when he got back from vacation, Barnes absorbed all he could about aircraft temperature control. After researching companies in the business, he and a partner bought a small firm that made insulation products for recreational aircraft. Thus, Thermal Structures was born.

The company he'd grown so effectively was Barnes's baby, and the team he'd assembled was made up of his friends. In meeting after meeting with suitors, eighteen in all, he became uneasy. Some clearly planned to streamline the company by cutting staff and operational expenses to make the balance sheet look better—before putting the companies on the market. Aggressive streamlining that often left targeted companies weakened and unable to compete in their respective markets had been a common occurrence since the late 1980s when a new kind of investor burst onto the scene—corporate raiders. Using a tool called the leveraged buyout, or LBO, they'd make unsolicited offers to target companies they thought were underperforming or had mediocre management. Or they would pounce because they saw that huge profits could be made on the deal itself no matter how the company fared after acquisition. Many LBOs ended up as train wrecks for the companies but often made the raiders rich. In a typical LBO, the targeted company—in this case, Thermal—would be acquired with vast sums of borrowed money collateralized by the assets of the company being taken over, which is why raiders loved the LBO. They typically put up little of their own funds in the game.

While the money offered by potential buyers sounded good, Barnes didn't like their answers to his concerns about his staff. "In every meeting I'd ask, 'What happens to my people?' And they would reply, 'Why do you care? You'll have your money,'" Barnes said.

Then Larry Mendelson arrived. HEICO's offer, though not as lucrative on paper as some of the others, had some very appealing sweeteners. Mendelson told Barnes, "We buy companies, we support them, and we grow them. We leave them alone, but if they ask for help, we help them." Barnes was also impressed that the Mendelsons wanted him to stay on. They and their co-investors offered $35 million in cash and HEICO stock in exchange for 80 percent of the company. Barnes would keep a 20 percent stake and remain the boss. Barnes had to think it over. He wasn't sure they would stay true to their commitment to being largely hands-off. In the end, though, the decision was a no-brainer. "Honestly, at first I didn't believe it," Barnes recalls about his doubts, "but they've done absolutely everything they said they would do."

One mark of Barnes's independence has been that he's been able to grow Thermal the same way that his now-parent company HEICO does: by making his own acquisitions with HEICO funding, following the same playbook. His most recent purchase was a specialty maker of industrial-scale generators that can power large construction sites or provide electricity to areas affected by storms or other natural disasters. Demand for these mega generators is at an all-time high. Barnes said he's never once been second-guessed by HEICO. "They'll ask me how they can help or if I need anything. That's just how they operate."

The acquisition of Thermal has been wildly profitable for HEICO. From sales of $17 million at the time HEICO acquired it, Thermal today has sales of almost $90 million. The company has grown to employ 400 workers and has opened locations in Plainfield, Indiana; Spokane, Washington; and Fargo, North Dakota. The 40-person engineering operation in Fargo exemplifies the economic boost that industrial-tech companies can provide for communities. While the average yearly wage in Fargo is around $30,000, HEICO engineers earn a starting salary of more than double that.

Going to Mars

Mergers and acquisitions are a common strategic tool in the Titanium Economy. But unlike sprawling conglomerates of the past that brought together unlikely enterprises with diversification in mind, industrial-tech companies look to deepen capabilities, become capable members of the supply chain, and lead their segments.

Arguably no other player does acquisitions better than HEICO, which, under the Mendelsons, has been rocket fuel for growth. Since 1990, HEICO has completed approximately eighty-two corporate acquisitions, achieving strong revenue growth and increasing cash flow—HEICO's secret elixir. Operating cash flow grew from $2 million in 1990 to $409 million in 2020, and cumulative free cash flow compounded from $1 million in 1990 to $2.7 billion in 2020.

Operating cash flow is the amount of cash brought in from a business's operations. Operating cash flow provides an important benchmark to determine the financial success of a company's core business activities. It indicates whether a business has enough cash on hand to grow without external financing. A company's free cash flow—what it must use for M&A and other development—is operating cash flow less capital expenditures. Cumulative cash flow is a term for projects or a company. Cumulative cash flow is calculated by adding all of the cash flows from the inception of a company or project. For example, if a company began operating three years ago, the cumulative cash flow for the company is the sum of what it generated in the three years.

Having good cash flow means the company boasts a perennially healthy balance sheet—attractive to Wall Street and the company's institutional investors alike—and also has the ability to fund its acquisitions at comparatively low cost instead of piling on debt. "The basic concept has been the same from the beginning," Mendelson said. "We look for high margins and high cash flow companies to buy."

HEICO looks for well-managed, entrepreneurial companies playing in high-margin niche markets that may be difficult to enter, such as companies that already carry needed certifications. The vast majority, like Thermal, are planted in American communities where they bolster local economies and help knit the civic and social fabrics of their towns and states.

The myriad companies that make up HEICO—such as Aero Design in Mount Juliet, Tennessee, Turbine Kinetics in Glastonbury, Connecticut, and LPI Corporation in Hollywood, Florida—don't tend to make headlines. Many of them are so specialized, serving niche segments in the precision avionics industry sector, that they have no need to seek publicity. But if you step back and look at them in aggregate, the collective innovation is staggering.

Consider this: The six-month, 293-million-mile journey of NASA's Perseverance rover, the fifth and most technologically advanced rover yet, which touched down on Mars in February 2021, was supported by crucial equipment produced by four different HEICO companies. One of those is VPT Inc., which supplied the converters and filters that help power Perseverance's camera system and onboard processing system. HEICO's 2010 acquisition of VPT, a Blacksburg, Virginia, maker of power modules for the space industry, perfectly exemplifies not only its uncanny growth strategy but the Mendelsons' bottom-line commitment to investing in companies that are lynchpins to thriving communities.

VPT is located in a high-tech corporate park that's just a few minutes away from the ridges, rivers, and forested ravines of the New River Valley and one of the most challenging stretches of the Appalachian Trail. CEO Dan Sable cofounded VPT with partners who were investors but who had no interest in running or working in the company. After twenty-eight years, they wanted to cash out, and Sable knew there was only one option. He went looking for a buyer.

Sable struck what seemed like a lucrative deal, but not with HEICO, which, he readily admits, he'd never heard of when the company lined up with other bidders. While Sable expected any buyout firm to improve operations and cut costs, when at the last minute the executives of the buyout firm tried to impose new conditions that would have resulted in massive layoffs, Sable got cold feet and the deal fell apart. HEICO quickly re-entered the picture and, while it didn't increase its original offer, Larry Mendelson startled Sable by telling him, "Dan, it would be hubris of me to say I could run your business better than you do." Indeed, VPT had been growing at an annual rate of 15–20 percent.

They struck a deal.

VPT's contributions to space exploration don't get major press for bold ambition as those of SpaceX and Blue Origin do, but VPT shoots for the stars in its own way. "We have three countries visiting Mars with satellites at present: the Emirates, China, and the US. And we're the only company in the world that has equipment aboard all three," boasted Sable, who has continued to lead the company since the sale to HEICO. Again proving the wisdom of HEICO's strategy to let good managers operate and get out of the way, he's maintained the company's impressive rate of growth.

By preventing substantial layoffs, Sable's choice to sell to HEICO demonstrates that success in the industrial sector by no means requires gutting American communities of high-quality jobs. VPT's home base of Blacksburg, Virginia, represents a possible future for many once-thriving American communities that have been hit hard by the transfer of so much manufacturing overseas, including smaller cities and rural townships. Blacksburg's population is just 45,000, but with the anchor of a high-quality university, Virginia Tech, providing good jobs as the town's largest employer—as well as a recurring stream of graduates with the technological skills needed for industrial-tech innovation, coupled with the synergy of technical expertise and accelerator services offered by the university—the town was well-suited for the creation of industrial parks, boasting three of them. The one in which VPT is located, the Virginia Tech Corporate Research Center, was established in 1985, a time of growing job loss in the industrial sector that had lost almost one-tenth of all jobs from a decade earlier. A for-profit, wholly owned subsidiary of the university, the center hosts 750 companies in industries such as aerospace, biotechnology, materials science, and electronics, among others. The town also features excellent public schools, a thriving performing arts community, and abundant opportunities for outdoor recreation, all of which contributed to *Businessweek* naming it the best place in the United States to raise kids.

Dan Sable has himself contributed to the synergy between Virginia Tech and the local employers, serving as an adjunct professor at the university's engineering school for the past nine years and teaching a class each semester in advanced electronics. When he approached the Mendelsons about the idea of teaching, which required him to cut back somewhat on his management duties, they were fully supportive. There has been one very beneficial result for VPT: over the years, Sable has hired scores of his engineering students. As we'll explore more in the next chapter, not only are quite a few other such Titanium Economy hubs thriving around the United States, but there is great potential for fostering many more of them all around the country.

From his office next to the sprawling, leafy Virginia Tech campus, Sable reflects on what might have happened to Blacksburg had its major employers closed or moved elsewhere as so many did in what came to be known as the country's Rust Belt.

The Playbook

HEICO and its portfolio of companies offer a strategic playbook that can serve as a model for Titanium Economy enterprises growing large—really large. HEICO's revenues hit $1.8

billion in 2020, with a net income of $314 million and a market cap of almost $17 billion. Not too shabby for a company that had sales of under $27 million when the Mendelsons took over. Listed on the New York Stock Exchange, HEICO has performed as well as FAANG in the past decade.

Since 1990, when the Mendelsons took over the Hollywood, Florida–based business, HEICO has generated a total return of some 47,500 percent. (No, that's not a typo.) During the past decade, its shares have increased by 1,270 percent, soaring past the S&P 500's 250 percent and even Berkshire Hathaway's 240 percent. Revenue reached $2.1 billion with a profit of $328 million in fiscal 2019—both record highs.

The key, according to Mendelson, is that the company sets a target margin of about 25 percent to support growth through acquisitions. By comparison, most other industrial-tech companies operate on margins of about 7–11 percent. In other words, HEICO earns a much higher return of profit on its operating expenses. Consider that if a company has a 7 percent operating margin, that means that for a product it sells for $100 it must invest $93, leaving it with $7—a ratio of thirteen to one. At HEICO, with its 25 percent operating margin, it must invest only $75 for a product that sells for $100, leaving it with $25. That's a ratio of three to one.

Fittingly, the Mendelsons have a lot of skin in their own game. With a 17 percent ownership of HEICO shares, they've amassed collective personal fortunes that *Forbes* magazine recently valued at $1.2 billion. They live well, but unostentatiously, in South Florida and don't discuss their wealth. They are happy, however, to discuss how HEICO's generous 401(k) plans, aided by the company's phenomenal stock market performance, have made millionaires of not just HEICO's team leaders but sizable numbers of machine operators. "They understand they're doing well because we're doing well," Larry said. "I take pride that employees think of it as their company too."

You'll get no quarrel on that front from HEICO managers and workers. Vaughn Barnes joked that if you walk around Thermal's offices on any given day, you'll find workers with one unerring eye on the task at hand and the other checking HEICO's stock price. "Everybody understands the connection between their efforts and the company's success," he said.

HEICO has 6,000 well-paid employees working at sixty-seven facilities in fifty US cities and towns. These workers get great benefits and on average earn three to seven times the salaries earned in the service sector—some in places where the poverty rate can exceed 15 percent. Many of HEICO's long-serving hourly factory hands have 401(k) retirement plans worth north of $1 million, and a handful of HEICO employees, including some long-serving secretaries, have amassed substantial retirement accounts with high returns. This is a remarkable stat, considering the median account balance at Vanguard for people ages 55–64 was $85,000 in 2020.

Many of its workers have stayed with the company for long careers, such as Judy Vetter, a former executive assistant to Larry Mendelson. She spent almost four decades with HEICO before retiring in 2019, and she counts herself lucky to be among the people who have been employed there. Vetter grew up in Miami, finished high school, and got married. When she divorced and desperately needed a job, she signed up for secretarial school. It was there that a classmate told her about an open position.

"I went for an interview thinking I could find a better job in a couple of years," she recalled. "I never got bored. I stayed for thirty-nine years." Vetter was able to send her kids to college without any of them needing student loans—something she and her coworkers hadn't expected they'd be able to do when they started in their field—and she now enjoys a comfortable retirement. "My husband is happy, my kids started their careers with no debt, and my grandkids will be taken care of," she said with a broad smile.

Recalling a ceremony a few years ago that was attended by Larry Mendelson, his sons, and about 200 employees honoring a HEICO maintenance supervisor who was up for his thirty-five-year pin, Vetter underscored how appreciated the quality of the Mendelsons' leadership is. "The guy was so happy that he ran up and kissed Larry in front of everyone."

HEICO has shown that while others packed up for China, Vietnam, Mexico, and the like, looking for cheap labor and relaxed regulation, American manufacturing companies could remain here and thrive. HEICO has expanded overseas, with global operations in Britain, France, and Asia, but it has never once considered abandoning the United States.

"There's a lot of talent here in the United States," Larry Mendelson has said thousands of times to his sons and managers. "You just have to know where and how to look."

HEICO is an industrial company that has broken out. It's a great success story, but it is not a one-off. *Theirs is a replicable strategy.*

On any earnings call with Larry Mendelson, all he talks about is cash, cash, cash. Not only is cash flow the language that investors want to hear, but cash flow allows HEICO to use M&A strategically to build a platform and scale within the micro-verticals they really care about. Micro-verticals, which we discuss in the next chapter, are a key feature of the Titanium Economy. Even though HEICO serves the aerospace industry, you're not going to find it providing the nuts and bolts of building an engine because it knows that 1,000 other suppliers can do it better, and it doesn't have a compelling value proposition. But HEICO will compete on spare parts for the engine with only a few qualified suppliers, and the company can lead a micro-vertical niche by acquiring smaller enterprises like Thermal Dynamics and VPT that provide excellent value in that space.

In addition to leveraging mergers and acquisitions, successful Titanium companies like HEICO employ two additional strategies: transforming the core and focusing on multiple expansion. The goal when transforming the core is to achieve better margins by leveraging data and technology to benefit the company. This ultimately leads to better, smarter products and more agile methods of decision-making—all resulting in better products for customers, a better customer experience, and more efficiency in operations and supply chain to serve customers. The net result is an enhanced profitability and cash-generation profile for the business.

The other strategy is for industrial companies to get the market valuation they deserve. Our analysis shows that core performance only explains a small portion of the actual multiple value that a company receives. In the past ten years, multiple expansions have become the primary engine of shareholder value creation, contributing as much as 90 percent of total shareholder value creation.

For those of you wondering what multiple expansion is and how it works, think of it like buying a house valued at $100,000. It needs some work, so you paint the walls, put in new appliances, and repair the roof. In other words, you fix the core. Then you put it back on the market, where it's valued at $200,000. Congrats: You just doubled your multiple. Let's say you believe it's worth even more and hire a staging company and a professional photographer while at the same time writing a blog on your new "up-and-coming" neighborhood's amenities such as parks, coffee shops, and a dining scene. You put the house back on the market, and voila: The market now values your home at $300,000, expanding your multiple three times over. Not bad for simply taking the time to tell your story.

Many companies in the Titanium Economy have not taken the time to tell their story. Furthermore, like remodeling the kitchen and bathrooms, there are specific things that hold more sway over multiple expansion—namely, the quality of revenue and corporate oversight. When it comes to quality of revenue, the *kind* of sales matters. Our research shows companies that have strong recurring sources of revenue are valued at as much as seven times that of one-off revenue sources. Therefore, if your product (one-off) can come with a sensor that provides customers with ongoing data about their own product and benchmarks with other users, and you can charge even a small premium for it, then the recurring revenue stream means exponentially more to investors.

In terms of governance, diverse board compositions, higher analyst coverage, and changing shares of low-turnover investor profiles all contribute to multiple expansion. The work doesn't stop there. As companies make changes to improve their quality of revenue and corporate oversight, they then need to ensure that the story is clearly explained to the market so investors understand the improvements that have been made. This is an area where many industrial companies still fall short.

With the combination of a programmatic approach to M&A like that of HEICO, a committed focus on multiple expansion, and investment in core business operations using the latest digital technologies, any industrial company can achieve the next level of performance in the Titanium Economy. Many companies can, and are, taking the HEICO playbook and tailoring it to their own situation to achieve the next level of financial performance and create franchises that optimize value for customers, shareholders, employees, and the communities they serve.

Chapter 5

The Power of Micro-Verticals

Shortly after William "Bill" Johnson took the reins as CEO of Welbilt in 2018, he asked to meet with Rick Caron, the company's head of engineering and resident genius. Johnson had high expectations for the meeting. Caron, an MIT graduate, had armfuls of patents to his name and was respected by people at all levels of the industry. He didn't disappoint. "In the first ten minutes of talking to him, I could see he was a brilliant man who had a passion for ideas," said Johnson, who began his career as a commissioned officer and nuclear engineer in the US Navy. "It was obvious that he was the perfect person to lead us on innovation."

Welbilt, a leading manufacturer of kitchen equipment, was spun off the Manitowoc Cranes Company in 2015, carrying heavy corporate debt from its former parent. When Johnson took over as CEO, the Welbilt board—including Carl Icahn, Welbilt's largest shareholder with an 8.4 percent stake—gave him a clear mandate: shed costs, accelerate technological advances, and remake the company into a leaner, nimbler, and faster player. To accomplish this, Johnson realized he needed to pry Caron away from nitty-gritty matters that consumed much of his time.

"He had a list of priorities that was ten things long," Johnson remembered. He told Caron, "Rick, I want to move you out of engineering. I want you to be the champion of innovation and digital for this industry."

Caron said that he wasn't sure he'd be happy leaving engineering. "These had been his people for a long time," Johnson explained. "That's when I looked at him and asked, 'Rick, do you really want to be dealing with people problems all day long, or do you want to spend your days innovating and saving this company?'"

That did the trick. "What do you need me to do?" Caron asked. Johnson didn't hesitate, responding, "Build me a common controller."

The mission was daunting, to say the least, but if Caron could pull it off, it would be game-changing. The idea came from a deep knowledge of one of the trickiest problems for restaurants and large food service providers, which causes the waste of masses of food and consumes precious staff time in a business with tight margins.

Welbilt is a leader in one of what we call the micro-verticals—small clusters of ten to fifteen firms in specialty market segments. The Titanium Economy is far from a monolith, with ninety micro-verticals. Micro-verticals have at least one of a homogenous set of customers, applications, or end markets. Serving a targeted, often small set of customers, each micro-vertical has its own characteristics, competitive dynamic, and structure and is grouped into a major segment. For example, laser devices and laser sensors are two of twenty-one micro-verticals related to electronic components and equipment. At any time, McKinsey tracks some 300 leaders across the spectrum of micro-verticals. The leaders are often seen as a "segment of one" since they lead and set the pace for the entire micro-vertical. Players in a micro-vertical compete, partner, and merge with one another, resulting in a sector that is "live and let live" compared to the tech world, which is much more "conquer the world."

Some micro-verticals are relatively new, such as electric vehicle battery producers, while others are in evergreen categories, such as industrial waste disposal, aircraft and aerospace parts and equipment, motors and generators, modular building materials, and industrial kitchen appliances, where Welbilt competes with a few firms for leadership. While each micro-vertical might be limited in scope, the size of each market is anything but small. The industrial kitchen equipment market is valued at $80 billion. The biggest player, Illinois-based Middleby, together with Welbilt, exemplifies the advantages of having granular expertise about customer needs, which drive all innovation.

Cracking the Black Box

Walk into any commercial kitchen and you'll find an arcade of equipment, from big-ticket items like cooktops, ovens, refrigerators, walk-in freezers, and HVAC systems to specialized appliances like espresso machines, ice makers, and deep fryers. Each piece of equipment requires close monitoring and regular maintenance. For example, if a refrigerator is just a couple of degrees too warm, food spoils. But if the temperature dips in the other direction, ingredients lose their nutritional value and flavor. What's more, in the clamor to serve guests, employees may not notice such slight temperature shifts, which can result in tossing out thousands of dollars of food or serving spoiled food, badly damaging an establishment's reputation.

For this reason, many restaurants tap a team member to check on such key equipment multiple times a day, which is a time-consuming means of monitoring that is also prone to human error. Johnson's idea would require utilizing cutting-edge smart technology to automate the process. "The journey we set out on," he explained, "was to develop a common controller and user interface for all of the equipment we supply to a kitchen."

Due to the depth of knowledge Johnson and Caron had developed about the food service industry—Johnson having formerly served as an executive with Dover, a manufacturer of refrigeration systems and food equipment—they understood that the rewards of such an interface would be tremendous. Workers would be able to track all equipment at a glance, and the system could be accessed from management offices or from workers' homes. The sensors could proactively raise an alarm when a reading fell outside the acceptable range. And with access to equipment status reports and alert logs, there would always be a record of readings for safety inspections and audits. What's more, a common tracker would give cooks more control. Once exact cook times and temperatures for a certain item were discerned—like those needed for getting just the right crunch in those French fries we all love—the information could be programmed across all the deep fryers in a chain, ensuring the same quality across the myriad shifts of workers and even locations.

In short, Caron had to create, as Johnson put it, "a world-class connected solution." The project exemplifies the opportunities the Titanium Economy offers for highly skilled engineers and designers to apply their skills to groundbreaking work. Caron had to hire a new team of people with the right expertise in creating interfaces and employing smart technology. Even so, whether the team could pull the feat off was very much in question.

They had to re-engineer every piece of equipment to speak a common language and work within a centralized information architecture, and each of those products was highly specialized, with its own sophisticated technology. These weren't just simple appliances that a part or two could be swapped out on to bring them under common control.

Welbilt produces a remarkable range of top performance equipment, having come a long way from its founding in 1929 by brothers Henry and Alexander Hirsch as the Welbilt Stove Company. Then based in New York City, the company is now headquartered near Tampa, Florida, in the small, laid-back town of New Port Richey, with operations still in New York as well as in Cleveland, Ohio, Shreveport, Louisiana, and Mount Pleasant, Michigan. As with so many hometowns of Titanium Economy companies, New Port Richey is a wonderful place to live, with a quaint main street featuring a nice array of shops and restaurants, as well as beautiful beaches and parks. Employees generally have plenty of time to enjoy the town's offerings, usually getting home from work in time for dinner. Caron's team working on the controller was an exception.

The Welbilt executive team had designated the project the company's number one priority, and Caron was given a tight timeline, so the team racked up overtime working out how to get all those highly differentiated pieces of machinery to work together. "Think back to the first refrigerator you owned," Johnson recalled of the task that faced Caron and his team, highlighting how much more sophisticated refrigerators are now. "Today's units run at a fraction of the energy and cost of that unit, and it's the same with our technology in fryers, grills, and cold units." For example, the company's Merrychef line features high-speed ovens with ultra-short cook times and ventless operation. Its Frymaster offers fully automatic filtration and oil quality sensors. Its pasta makers have electronic, programmable controls. Welbilt's Garland countertop ranges, griddles, and broilers use induction technology for precise temperature control. And its Lincoln ovens deploy air impingement technology for rapid heating, cooking, baking, and crisping, so pizzas come out of the oven perfectly crusty every time.

Caron's team had to re-engineer one machine at a time to function with a new controller, and the controller had to integrate every machine into a single functioning system. All the machines had to speak to one another, and they had to factor in Welbilt's three services: KitchenCare, for aftermarket parts and service; FitKitchen, for fully integrated kitchen systems; and KitchenConnect, the cloud-based digital platform solution.

While the FAANG tech stars have dominated media coverage of the advent of the long-hyped, but now rapidly developing, internet of things (IoT)—offering consumers smart doorbells, smoke alarms, and vacuums—industrial-tech firms have been seizing opportunities to bring smart technology into both their manufacturing processes and their products. Today, every pump, motor, compressor, conveyor, or bolt can generate data, and breakdowns can be diagnosed with a simple click of a button, proactively determining what maintenance needs to be performed and replacing parts before they even break. Equipment can also be controlled with exquisite precision, and rich troves of data can be collected and analyzed to spot opportunities for performance enhancement. Welbilt's aim was to enable its products to collect data regarding temperature, flow, pressure, and humidity so its clients would get a real-time snapshot of their equipment in a single user-friendly interface.

The smart technology was readily available—touch screens, sensors, and QR codes paired with the latest custom software. The harder part of the job was that Caron and his team had to start from the ground up to build a common infrastructure for a proprietary controller system, or the physical touch screen interface powered by custom software to control the kitchen systems. This meant installing the same basic hardware across the entire product line, requiring enough computer chips, sensors, and other materials for millions of units. These changes needed to be made without diminishing the integrity of each product's unique functionality and specs—many of which Caron had developed in his previous role as head of engineering.

After three years, the task was complete. In 2021, Welbilt introduced a fully integrated line of products, which all operated through a proprietary common controller touch screen featuring a user-friendly dashboard under the slogan "Bringing Innovation to the Table."

Rick Caron died unexpectedly in 2021, but not before he saw how his common controller innovation revolutionized the world of commercial appliances and was delivering true value for all stakeholders.

"The common controller was an important next step in modernizing our industry," Johnson said, "but the unintended benefit has been in helping to alleviate supply chain shortages." Not only could they stock up on needed parts in case of delay, but because all products used the same exact controller, whether an oven or refrigerator, they could warehouse them from a central location and move them across businesses and different platforms as demand dictated. "The strategy saves money because now all our machines are uniform."

The uniformity had the added bonus of enabling Welbilt to scale faster when needed. The company easily ramped up production to meet demand. Customers bought products knowing the same controller worked without having to retrain staff. The user-friendly interface made training easier and more uniform, cutting costs across the board for Welbilt and its customers. Software upgrades were applied across the various product lines, and an improvement for the brand extended to all. "Now, all of our equipment is born digital and can easily be connected to any department in Welbilt," Johnson added. It also facilitated warranty claims and sales support.

The innovation is transforming the company: In 2021, net sales were up 92 percent from the prior year, with an adjusted EBITDA nearly doubling from 9.6 percent to 18.6 percent. Net earnings were $31.6 million compared to a net loss of $32.5 million in the prior year. A good part of the turnaround can be attributed to the common controller, which made customers more likely to turn to the company's other lines rather than mixing and matching equipment. Their efforts have been so successful that they emerged as a segment of one in the space, and in 2021 Ali Group acquired Welbilt in an all-cash transaction for $3.5 billion in aggregate equity value and $4.8 billion in enterprise value to create an industry-leading pair of food service equipment providers.

In the Titanium Economy, there's new wealth to be found in data. Leaders will be those who can aggregate, cleanse, link, and monetize the data flowing through their enterprise. Welbilt did just that by bringing its product lines under one common architecture. But integration can mean something entirely different.

Refreshing the Air

So much attention regarding innovation has been paid to industry disruptors who move into an established terrain from outside and upend legacy businesses with fresh perspectives. In Titanium Economy companies drawing on micro-vertical expertise, great innovation like that of Welbilt is being achieved by insiders who are willing, and able, to spot the possibilities for applying new technologies to disrupt their own business. They offer abundant opportunities for engineers, product designers, and operations and supply chain specialists to work with the most advanced new technologies to find novel ways of bringing value to customers. As a result, they sometimes solve long-standing problems in their vertical and end up dramatically enhancing their customers' operations and results in the process.

CaptiveAire has done this to capitalize on an opportunity in the ventilation systems niche of the industrial kitchen micro-vertical, becoming the industry leader in their niche in the United States. With operations in Iowa, Oklahoma, California, Florida, and Pennsylvania in addition to its home base of North Carolina, the company employs 1,200 workers and earns over $500 million in annual revenues.

"Let's face it: computer boards are running pretty much everything," said Robert Luddy, the founder and CEO of CaptiveAire. "With everything running on the internet, and with IT systems so crucially important, we've become a very high-technology company." Luddy carries himself with the posture and direct manner of the former Army E-5 that he once was. He challenges his engineers to constantly propose new ideas for innovation and has fostered a culture at CaptiveAire in which the best idea wins, no matter who thought of it, and is then tested, and tested again, and is either adopted or the team moves on to a new idea.

Luddy started the company that's become CaptiveAire in a one-room workshop in Raleigh, North Carolina, in 1978 with $1,300. Originally called Atlantic Fire Systems, the company's specialty was making fire suppression systems for restaurants. Demonstrating Luddy's prowess for spotting niche opportunities, in the first year the company earned $297,000, and by 1979, revenue grew to $1 million.

In working with restaurants, Luddy noted another important need that could be addressed. The thing about industrial kitchen ventilation is that every cooking facility needs one, but until Luddy came along, most systems hadn't been updated in decades. He pointed out, "In simple terms, restaurant kitchens were hot, greasy, and uncomfortable. Kitchen ventilation was a slow-moving industry at that time, with many products below the threshold of quality needed for modern restaurants."

In the latter half of the twentieth century, the system most restaurants had in place had worked essentially the same way for decades: an exhaust hood captured cooking fumes, hot air, grease, and smoke and moved them along a series of ducts in the ceiling up and out of the building through an exhaust fan that was typically on the roof. At the same time, what's called a make-up air unit, also typically located on the roof, pumped outside air into the building to replace what was taken out and helped maintain proper air flow and building pressure. But these traditional ventilation systems not only recirculated viruses, odors, and other contaminants floating in the air, but they were also inherently inefficient since different manufacturers produced the various parts used in making them. So when the old vent system would break down, repairs were difficult to diagnose and expensive to fix.

Luddy began improving ventilation technology by integrating fire suppression technology into kitchen ventilation hoods.

He kept developing from there, discovering lots of room for improvement. Recalling the sorry state of the kitchen hood system when he entered the business, he explained, "Exhaust flow rates were high, consuming too much energy; kitchen hoods were not very reliable; and the available systems did not provide much creature comfort." So he charged his team with reinventing the traditional ventilation system to circulate air at a far higher, more efficient level in terms of cubic feet per minute while doing a better job of containing irritants.

Most recently, the team took on the creation of a transformative new system to replace the long-dominant make-up air equipment with more advanced engineering, including smart technology. As John Hess, a mechanical engineer in the company's R&D division, said to us, "Just because everything has been done the same way for the last ten, fifteen years—even longer than that—doesn't mean it's the correct way. You always have to be innovative. You have to be open to using those new technologies that are available to you." This embrace of change, even to products the company had made the backbone of its success, is a stellar example of how limber many Titanium Economy companies are, breaking out of legacy thinking. Bill Griffin, president of engineering and manufacturing at CaptiveAire, highlighted, "We've been promoting dedicated make-up air for over twenty years." But by keeping well abreast of the wealth of advancements in technology in recent years, he and his people were able to perceive that they could vastly improve their own offerings.

"We took a step back," Griffin explained, "and saw the kitchens had multiple units on the roof doing the same thing.

Our idea was to eliminate the make-up air unit, eliminate the make-up air plenum, eliminate the return duct in the kitchen and the RTUs, and replace all of that overlapping technology with a unit in the kitchen and another unit in the dining room."

CaptiveAire's "Dedicated Outdoor Air System," or DOAS for short, had immediate benefits. First, because the system had fewer processes involved—only one DOAS per room—the duct system was much simpler. CaptiveAire's redesigned ducts eliminated the need for drop ceilings, the bane of any restaurateur. Plus, lower weight on the roof alleviated structural stress. Both of these aspects drastically reduced installation costs and made repairs both cheaper and easier. On top of this, because the DOAS replaced the make-up air system in the hood, turbulence was reduced, decreasing the amount of energy used. And with more fresh air pumped into the kitchen, temperatures became better-controlled and the need for recirculated air was reduced.

"It was a major change in philosophy," Griffin said. "Instead of focusing on just the hood, or just the make-up air, every piece of ventilation was part of integration." So while restaurants no longer have to hide bulky, extraneous ducts in drop ceilings, the benefits were more than cosmetic. The simpler system required fewer processes, meaning fewer things to break down.

"DOAS has caused one of the biggest changes in the restaurant ventilation industry that I can remember," Griffin said. Instead of getting parts from different manufacturers, CaptiveAire decided to keep it all in-house. "We make all the products. We make sure they're compatible and sustainable. It took a long time to do that," he added.

Integration is a recurring theme in Titanium Economy innovation, and it's capitalized on in various ways. Both CaptiveAire and Welbilt have taken advantage of the ability to create newly integrated equipment. Another major innovator in this micro-vertical, Middleby Corporation, based in Elgin, Illinois, has become the largest manufacturer of commercial cooking equipment in the world through a strategy of vertical integration achieved by the shrewd acquisition of companies based on its depth of industry understanding. In the process, Middleby has greatly enhanced its technology capabilities.

Under One Roof

One of those industrial giants you've likely never heard of, Middleby's body of companies, numbering ninety and counting, covers the breadth of food preparation and service. This integration allowed the company to establish itself as a segment of one in its micro-vertical. The company makes everything from the Viking ranges coveted by home bakers to conveyor belt ovens for food producers. Its uncanny eye for innovation had enabled it to snap up hot players, making fifty acquisitions from 2010 to 2019 that have allowed it to stack a fortress of products and companies under one literal roof.

If you want to see what the future of food service looks like, fly into Dallas's Love Field Airport and drive fifteen minutes down the Sam Rayburn Tollway, taking the exit at Standridge Drive into the unassuming corporate park complex. There, among the colony of boxlike buildings, you'll find a chef's theme park of kitchen equipment in a space that's nearly the size of a football field.

Anyone can visit the Middleby Innovation Kitchen, or MIK. On display are over 150 food and beverage preparation tools offered by Middleby. Along one wall are banks of gleaming stoves. Along another, granite countertops with table settings of white plates and wine glasses surround collections of the latest high-end kitchen equipment. In an adjacent corner, a blue-tiled firepit stands by two tandoor ovens.

The space has become a sort of Disneyland for food professionals. Chefs Middleby has on hand will cook, serve, and clear your plate. Better still, whatever tech you want to try for yourself, your restaurant, your distribution service, or your food brand is ready to be taken for a spin. Restaurant and quick-service food operators can explore the equipment and, with the expansive display space, see how various pieces might fit together and look in their kitchens or food prep areas.

Middleby Master Chef Russell Scott, the guru in charge of the MIK, explains that the company perceived it could significantly improve the ways customers were shown new products. "We reviewed what was taking place at food convention shows and asked ourselves, 'How can we improve on that?' You can come here and work on a particular need or area where you want to expand, whether that is in pizza ovens, sous vide, beverages, or various refrigeration tech. We cover every application and type of cooking you can imagine. Here, we do a deep dive into all things hospitality.

"The kid in the candy store analogies work for me," he said. "Every day, here we are, working and learning what we can by utilizing these different pieces of equipment."

Under the leadership of Chairman and CEO Selim Bassoul, who left the company in 2019, and its current CEO, Timothy Fitzgerald, Middleby has a standout record of identifying sweet spots of new customer behavior and applications of new technology. The company's acquisitions of grill manufacturers Kamado Joe and Masterbuilt, for instance, have allowed it to capitalize on the substantial upswing in people cooking outdoors in their backyards, which began before the Covid-19 pandemic but was also greatly accelerated by it. Overall sales of gas grills, for example, grew 37 percent in 2020. Similarly, with automation overtaking food processing, the 2018 acquisition of Castelnuovo Rangone, Italy, based Ve.Ma.C. has been a feather in Middleby's cap. With $15 million in annual revenue, the company designs and manufactures robotics solutions for protein food processing lines.

Middleby's deep expertise in the segment has driven this ability to occupy profitable niches and get ahead of the competitive curve. It leapt at the chance to serve the trend of ghost kitchens during the Covid-19 pandemic to meet increased demand for food delivery, which gave rise to independent chefs, food brands, and small vendors. Think WeWork for cooks. Middleby ghost kitchens feature an array of commercial-grade appliances in one site to facilitate food prep and easy third-party pickup from Uber Eats, Grubhub, and legions of other delivery services.

Knowing the pressure points for its customers, combined with a corporate ethos that "no problem is too small," has allowed the company to seize opportunities like that presented by Skyflo. This is a smart technology made possible with products created by L2F, a manufacturer of robotic products for kitchens that Middleby acquired in 2018. As a wireless control device attached to wine and liquor bottles, Skyflo helps servers pour just the right amount every time, with no under- or overfilling, and as a result, operators of bars, restaurants, concert halls, and stadiums can track sales and inventory down to the pour. As one equity analyst comments about this niche integration: "Even with something as simple as ice cream machines, they are able to create a higher-value product through their internal levers, improve the margins on their acquisitions, and grow their relationships with customers through another valued product offer."

Middleby, like HEICO and other superstars in the Titanium Economy, follows the playbook of creating greater value through technology and leveraging additional earnings toward synergistic acquisitions that allow it to lead its micro-vertical.

As Middleby adds more companies and capabilities to its portfolio, MIK brings them together for customers, showcasing ever more eye-popping technological advancements. And that no longer requires a trip to Dallas. Today, a network of eighteen Middleby Test Kitchens covers the country nationwide, reaching potential new and existing customers in every facet of the industry, from fast food chains to hotels.

The company has racked up annual sales approaching $3 billion, bolstered by a 16 percent CAGR from 2010 to 2020, with 84 percent organic and 16 percent inorganic sales from acquisitions in the fiscal year 2020. And even amid the ongoing uncertainty over Covid-19, Middleby's net sales shot up by more than 70 percent in the second quarter of 2021, with nearly 72 percent year-over-year growth in comparison to 2021. In other words, if you invested $1,000 in Middleby in 2005, you would have had $100,000 by 2021—a better return than if you'd put that money in several of the FAANG companies.

While thousands of tech startups never make it through their first year in the current kill-or-be-killed business environment because they can't deliver on their business plan or move quickly or nimbly enough, the expertise developed by micro-vertical industrial-tech firms allows them to chart solid strategic moves into competitive white space. Many also inhabit niches in live-and-let-live ecosystems and even in micro-verticals with considerable competition among segment leaders, as is true for Welbilt and Middleby. Many of these micro-verticals offer plenty of room for growth. Welbilt and Middleby, who are two of the biggest publicly traded players, together control just 10 percent of their overall segment. Aspiring innovators looking for exciting opportunities to apply their engineering and design skills should turn their attention to these micro-verticals; investors should also take note. The application of the panoply of new smart technologies to breakthrough solutions in these segments has only just begun.

Chapter 6

The Great Amplification Cycle

The conventional wisdom that American manufacturing jobs have streamed out of the country to Mexico and Asia since the 1970s isn't wrong. However, for every Flint, Michigan, that has been infamously devastated by the loss of jobs in the automotive industry, there's a Simpsonville, South Carolina. This once-prosperous mill town, population 25,000, located in Greenville County, is part of the hub of industrial tech known as the Golden Strip. Near the border between North and South Carolina, the Strip also includes the South Carolina towns of Mauldin and Fountain Inn and is home to industrial-tech firms such as Magna, Bosch Rexroth, and Mitsubishi Chemical.

Once a thriving center of textile and apparel manufacturing, the massive shift to Asia beginning in the 1960s might have left the area in a long-term economic depression considering how concentrated its economy was in the industry. Instead, the towns of the Strip benefited from a process of revitalization by which industrial-tech companies cluster in areas in what we now call the Great Amplification Cycle. Just as in the golden age of the US industrial sector between 1948 and 1973, when manufacturing hubs flourished in the heart of America, Titanium Economy hubs are booming all around the country. Their growth often gains momentum once an anchor company, or a small group of them, attracts increasing talent and investment, which in turn attracts additional companies and even more talent and investment in a flywheel of economic growth that's a hallmark of the Great Amplification Cycle.

For the towns of the Golden Strip, the flywheel was set in motion when the tire company Michelin, which had operated a single factory there since the 1970s, decided to locate its North American headquarters in Greenville County in the mid-1980s. Once Michelin established that beachhead, others took note, impressed by the area's business-friendly climate. Competitor Bridgestone followed, as well as the carmaker BMW, whose state-of-the-art production required plenty of engineers and skilled workers. Soon, factories in the area were cranking out not just high-end cars but also wide-body airplanes, medical devices, and advanced materials.

As was true for the car industry in Detroit and high tech in Northern California, the convergence of industrial, tech-oriented companies allows them to capitalize on synergies, which we also saw with the thriving cluster in Blacksburg,

Virginia, with universities and community colleges playing an important role, as was famously true for Silicon Valley. State and local governments, as well as think tanks and private institutions, fuel the process by establishing and promoting enterprise zones and hiring incentives. South Carolina also improved roads and other infrastructure to support these enterprises.

Titanium hubs exist in Dallas–Fort Worth and Houston in Texas, Chicago and Milwaukee in the Upper Midwest, Tampa and Miami in Florida, and Omaha, Nebraska, which has earned the nickname "the Silicon Prairie." But really, they're everywhere. A crucial difference between these Titanium Economy hubs and Detroit and Silicon Valley is an enhanced diversity of industry. In the greater Simpsonville area, alongside BMW and Michelin, you'll find tractor and farm equipment company Caterpillar, floor products manufacturer Techtronic Industries, and steel producer ThyssenKrupp, to name just a few from the range of sectors. That diversity insulates against the hollowing-out of communities that can happen if they are overly reliant on one industry or even one firm, protecting them in case that industry or that one factory goes into a downturn.

An example of that fate is Janesville, Wisconsin, which was made the focus of national media attention in part because of the book *Janesville.* The once-bustling city relied heavily on a General Motors plant, which for years provided residents with stable, gainful employment. In the 1970s, the plant employed as many as 7,100 people in jobs that paid well enough for employees to maintain comfortable middle-class lifestyles, allowing them to buy nice homes and provide their children with college educations. But as part of the decades of upheaval in the American automobile industry, in December 2008, the plant was closed.

For many Janesville residents, the loss was devastating. Grandparents, parents, aunts, uncles, and siblings in some families had all worked at the plant. But the ramifications of the closing were felt well beyond the families of employees, in the opposite of the Great Amplification Cycle. A locally owned manufacturer that made the seats for the cars produced at the GM plant soon found that without the plant as a customer, it was forced to close its factory, resulting in another 800 lost jobs. And as thousands of Janesville's families began cutting back on their expenses, small businesses all over the area struggled. Hair and nail salons saw huge reductions in appointments because longtime customers could no longer afford to go as frequently. Restaurants suffered as families chose to forgo their weekly dinner out. Realtors couldn't sell houses.

Perhaps most telling, however, was the sharp increase in the number of students eligible for free and reduced lunches, a metric commonly used for measuring poverty levels. In the early 2000s, one-third of Janesville's students were eligible for free or reduced lunches. Two years after the plant's closure, however, that number had risen to over half the student population; at three elementary schools, it climbed past 70 percent—statistics that have held steady in the years since.

The McKinsey Global Institute modeled where the United States is headed today at a granular level based on zip code, population demographics, and a whole array of details. Going purely on past data, we found growth in urban and suburban areas with tech- and service-related sectors, whereas local economies that once held more traditional manufacturing industries lost employment compared to their counterparts in the high-growth hubs. When we extend that data to project the future, we find the same recurring theme: industrial centers losing jobs to metropolitan hubs. Based on current projections, 25 percent of the urban areas will get 60 percent of the total job creation in all the United States. The gains in the future, if we stay on the current path, will be concentrated in a select set of communities.

Simpsonville, which had relied so heavily on cotton milling, could have easily followed a similar trajectory as Janesville. It's easy to appreciate why such concentration in one industry can occur when big booms bring great prosperity. In the decades following the Civil War, workers flocked to Simpsonville, and new neighborhoods sprang up to accommodate their families. The boom made building a trolley line to connect downtown Simpsonville with the cotton mills affordable. There were so many jobs, in fact, that people dropped out of high school to work at the mills because it paid too well *not to.* Anticipating the rapid decline of the US textile industry would have required a magical crystal ball. At the peak of the industry in the 1970s, upwards of 400 mills were thriving, concentrated in the South. The number dwindled over the years and by 2001, the industry had all but collapsed.

Many communities dependent on the industry suffered as Janesville did. But in South Carolina, then among the poorest states in the nation, with limited infrastructure and a struggling economy, civic leaders recognized that with sprawling acres of land available for building new types of warehouses and factories, they should woo companies to revitalize the economy. Public and private interests formed partnerships to improve infrastructure from ports to roads. South Carolina Governor Ernest Hollings in the early 1960s pushed for improved technical education in the state to attract new industry, resulting in the state's first technical college in 1962. Forty years later, when the big, marshmallow-like tire mascot came calling, the state was ready. "One of the key attributes that attracted Michelin to South Carolina more than forty-five years ago was the presence of a skilled and talented workforce," said Will Whitley, director of state and local government affairs and community relations of Michelin North America Inc.

Michelin now has 9,000 employees spread across fourteen sites in the state, with its headquarters in Greenville County. It's joined by nearly a dozen name-brand companies and manufacturing plants that operate in the Strip. Boeing, with 5,700 employees in the area, builds its 787 Dreamliner while Lockheed Martin makes fighter jets. BMW's plant in Greer cranks out more cars than any other BMW factory worldwide. Volvo opened a plant in 2018 to make midsized sedans. Mercedes is there, as is Samsung to manufacture home appliances.

With the influx of new workers spending their cash at local businesses, some smaller industrial-tech companies in Simpsonville benefited. One of those was Sealed Air.

When Plant Manager Donald Hebert turns off North Maple Street into the parking lot of the Sealed Air plant in Simpsonville in the morning, he often notices groups of employees who've just come off the plant's night shift, hanging in chatty clusters. For many in town, the plant serves not only as a place of work but as an unofficial community center. Over the holidays, the plant hosts parties to which anyone in the community is welcome. Employees make cookies, hold holiday-themed contests and giveaways, and Santa Claus listens to children's Christmas wishes. The come-one-come-all holiday party is just a small part of the role that Sealed Air plays in the civic life of Simpsonville and the surrounding area. Throughout the year, Sealed Air volunteers take shifts cleaning and painting the free medical clinics in the area that treat those who can't afford medical care, even bringing their own pressure washers and painting supplies. "The clinic staffs appreciate it," Hebert said, "because our folks make sure no one is left out."

Simpsonville offers plentiful jobs, overperforming schools, and better-than-average levels of healthcare. At 3.6 percent, it has one of the lowest unemployment rates in the state with a median household income of $71,990 and a median home price of $173,400. Meanwhile, the Greenville County School District boasted thirteen elementary schools in South Carolina's top fifty in 2018. Simpsonville also features a surprisingly vibrant cultural scene. Despite the city's relatively small population, it has a large outdoor concert venue at Heritage Park that attracts major artists like Santana, Counting Crows, and 3 Doors Down. All these enticements draw new residents. In fact, Simpsonville is currently one of the fastest-growing communities in South Carolina, named by Zippia as one of the state's best towns for working women. And HomeSnacks, the data-driven website that combines US Census data with dozens of other sources to put a spotlight on America's regional hidden gems has included Simpsonville in its "10 Best Places to Live in South Carolina" list every year since 2018.

Sealed Air is a lynchpin of the town's prosperity, its workers' pay filling the coffers of local businesses. The company has come a long way from its humble origins. In 1955, the two engineers who founded the firm set out to create an innovative type of wallpaper that would double as insulation. They failed to make their product a success as wallpaper, but their creation went on to become a commercial hit in another guise.

"Everyone knows us as the bubble wrap company. We invented it. If you think about it, chances are you have touched our product on a daily basis," said Ted Doheny, CEO of Sealed Air. "If you received an e-commerce package, it came with bubble wrap. If you went to your local supermarket to get meat, it was protected by our Cryovac."

Sealed Air now leads the packaging industry, producing packaging materials for food and consumer products made from a variety of advanced materials that offer innovative features. Doheny explained, "We are reinventing ourselves to become a world-class, digitally driven company automating sustainable packaging solutions. We are heavy into developing sustainable materials to make the transition to a plastic-free world and are leading from the front to eliminate plastic waste. In 2018, we signed up to make our packaging solutions to be 100 percent recyclable and reusable by 2025, and we are investing in blockchain to drive 100 percent traceability on our packages." Packaging meat products with a digitally scannable blockchain code enables tracking the chain of custody by producers, packagers, and distributors. This digital block protects against counterfeit meat and contains scannable consumer information about origin and nutrition to satisfy demanding consumers who want to know where their food comes from and if it's sustainable. "We are purpose-driven in everything we do: we are in the business to protect, to solve critical packaging challenges, and to make our world better than we find it," Doheny added.

Sealed Air's massive Simpsonville facility spans 1.4 million square feet of space and employs over 1,000 people, operating day and night, 365 days a year. Since 2018, Sealed Air has delivered sustained EBITDA improvements, expanding by 9 percent or more each quarter (equal to more than 400 basis points of margin expansion). As a result, the company stands out from its pack of peers and becomes what we consider to be a segment of one in margin performance.

As is true across the US manufacturing sector, one challenge for Sealed Air is finding the talent it needs. On any given day, half a million factory jobs remain unfilled in the country, holding back industrial growth. To address the problem, Sealed Air has developed good relationships with the area's many educational institutions, from the elementary school level up. The company has partnered with student councils at regional primary and secondary schools in a range of activities, assisting the students with projects from designing T-shirts to writing mission statements and developing leadership studies in a civic-minded pursuit that also carries an ulterior motive. "We help to educate the community in terms of food packaging," Hebert said of his Community Engagement Team. "We talk about how what we do goes beyond packaging, how nearly everything we do is about preserving shelf life, preventing food waste, innovating new ways to recycle plastics, and reducing greenhouse gases. We show them the technical side of the industry, which typically interests the students, and that allows us to talk about the career opportunities we offer in manufacturing."

In addition, Sealed Air assists Simpsonville's Center for Community Services with GED Bootcamp classes and job fairs and runs an eighteen-month employment program for students at Greenville Technical College and Spartanburg Community College. Students can earn eighteen dollars an hour working part-time, and the company picks up the tab for their tuition and books. These high schoolers are getting skills that can serve them in finding high-quality work to last a lifetime. Some also earn scholarships through programs like Apprenticeship Carolina, making technical training affordable. Such programs have helped South Carolina rank second nationally in the concentration of team assemblers; third in inspectors, testers, sorters, and samplers; and fourth in the number of industrial engineering technicians—thanks in large part to community colleges and technical schools.

With the state's two public research universities of Clemson and the University of South Carolina churning out graduates, Sealed Air and other Greenville County manufacturers work closely with them on their Tech Scholars initiative, both giving back to the community and helping to ensure a steady stream of hires. Clemson's Advanced Materials Center and the University of South Carolina's Center for Electrochemical

Engineering both push the envelope of research and innovation and foster a pipeline of skilled workers. How fitting that materials science has become a specialty in the area, as advanced materials are to the twenty-first century what textiles were to the twentieth century.

The human resources team at Sealed Air must compete for talent against bigger-name industrials in the vicinity, such as Michelin, GE, and BMW, but they more than hold their own. Not only are Sealed Air's pay and benefits highly competitive, but management and employees are eager to talk about the company's enviable culture—something Plant Manager Donald Hebert said goes much further than any brochure or hiring pitch ever could. That's been enhanced by Sealed Air's focus on diversity in hiring and promotion policies, resulting in a workforce that is 31.3 percent female and 40.9 percent ethnic minorities. As is a hallmark of many industrial-tech success stories, the company has been a magnet for new immigrant families. One hope in developing industrial-tech hubs is that the momentum of growth helps bring a healthy influx of both skilled workers and children who may grow up to become Golden Strip employees.

The Great Amplification Cycle just keeps turning, with enterprises of all stripes continuing to flock to Simpsonville. Among them are firms that naturally support industrials, such as Sunland Logistics, which provides warehousing, and Enviro-Master, a commercial cleaning business. But the boom is evident in the multiplier effect beyond the industrial base. Marco's Pizza, a national chain that already operates three other locations in the greater Greenville area, opened in a

Simpsonville shopping center alongside a Starbucks and a Publix supermarket. Ivybrook, a Montessori school with forty locations across the United States, is opening its fifth South Carolina location in Simpsonville to meet demand from the growing number of families with young children. Nail salons, yoga studios, and florists are popping up around town. There is no end to growth in sight.

A Thousand Teslas

The dynamism of hubs like the Golden Strip has been brought even to the long-beleaguered US automotive industry, thanks in part to Elon Musk. He has made the business sexy again by reinventing the automobile, among America's most mature and iconic consumer products. Tesla is doing what no one else in the world thought possible in the process. By producing innovative electric cars, not in a faraway, low-cost country, but in a technologically sophisticated factory in high-cost California, the brand is redefining what it means to be an industrial company. The company's successes have been well-documented. What's not been as well-documented, however, are the positive ripples of that success and the multiplier effects. Tesla has set in motion an enormously powerful Great Amplification Cycle.

Tesla is now the biggest manufacturer in the Silicon Valley area, bringing car manufacturing back to Fremont, California. Between its automotive operations and aerospace and alternative energy units, it provides 50,000 jobs and generates an estimated $5.5 billion in economic activity there. And it's expanding operations—not offshore, but in rural Nevada, Buffalo, New York, and Austin, Texas.

Setting aside the untold benefits to the vendors and small businesses in the communities based around Tesla plants, like those in and around Simpsonville, Tesla has made auto manufacturing attractive and highly valued again, talented professionals clamor to get Tesla on their resume, and investors are eager to enhance their portfolios with the company's stock.

Ah, but you say, "Musk is a one-off, a contrarian, a mad genius. Tesla is an aberration, a unicorn's unicorn." We disagree. The very premise of this book is that we are capable of producing a thousand companies akin to Tesla in every industry. Part of what Musk understood—something that has also been embraced by other industrial leaders—is that the core of the American spirit is about controlling one's destiny. And as the saying goes: Where there's a will, there's a way. Indeed, good manufacturing jobs are returning to the United States—a trend accelerated by the fragility of the global supply chain that was highlighted by the Covid-19 pandemic. The trend of reshoring workforces has accelerated. The United States is on track to add 220,000 jobs from reshoring or foreign direct investment in 2021, up from a record 160,000 jobs in 2020. Looking forward, reshoring is poised to surge 38 percent to a record high.

For a growing number of companies, the value of reducing labor costs by building things in other nations has diminished. Freight and insurance costs are skyrocketing, and lead times are increasing with more product delays. Meanwhile, labor costs in China are rising. Companies are looking to mirror or create redundant supply sources to strengthen supply chain resilience.

The Titanium Economy is contributing significantly to the regrowth of jobs. Manufacturing now accounts for 20 percent or more of employment in 460 counties across the United States. In fact, industrial employment shot up from 11.6 million in January 2011 to 12.5 million in October 2021—in the midst of the Covid-19 pandemic, no less. In just the food processing and manufacturing sectors that are home to Sealed Air, jobs grew from 1.46 million to 1.65 million in the same period.

To be sure, images of dire Rust Belt decay make for dramatic television and photographs, but they also feed a stereotype that manufacturing is dead—an assumption that is just flat-out wrong.

Still, while Titanium hubs can be fostered just about anywhere, they are far from random or spontaneous; building them requires the kind of strategic collaboration among corporate leaders, local government, and educational institutions that has propelled success in Simpsonville. The playbook has been established, and we say to communities all around the country: let the games begin.

As we look to the future, we envision towns of every size, stripe, and locale anchored by hubs of collaborative innovation where industry works hand-in-hand with academia and government to solve the myriad tough problems that are plaguing the United States and countries all around the globe. These range from issues of climate change to the need for better health technology, the challenges of decaying infrastructure, and a need for more innovation, which we'll explore in the context of family business in the next chapter.

If properly nurtured, Titanium hubs can ensure that value added through innovation will spread unabated throughout industrial-tech supply chains. Local synergies—accumulated region by region—will foster greater efficiency and resilience, and this will result in untold rewards for owners, workers, and local small businesses of all kinds. Crucially, it will also revitalize the nation's overall competitiveness.

Chapter 7

Titanium Family Values

In September 2021, Dot Foods CEO Joe Tracy sent out a video message to the drivers behind the wheel of the company's big rigs: "You guys are our last line of defense. During this crazy, complicated Covid environment, you've been there every day—traveling throughout the country, serving our customers, and doing a phenomenal job." Tracy wanted to make sure that Dot's ground army of drivers felt appreciated because, as Tracy sees it, what enables Dot Foods to be a powerhouse food company is its ability to attract and retain drivers better than anyone.

Tracy is one of twelve children of the family that in 1960 founded the Associated Dairy Products Company—later renamed Dot Foods after his mother, Dorothy, a cofounder. He grew up being trained to run the business. His father, Robert Tracy, started the company, selling powdered milk out of his station wagon and trucks. The company branched out into offering other dry products and then moved on to beverages before expanding geographically. In the more than half a century since those humble beginnings, Dot Foods has become one of America's largest food redistributors. This segment of the food industry involves buying food products from manufacturers that are selling relatively slowly or may be at risk of spoiling and warehousing them for sale to a range of customers, including other distributors and wholesalers.

Dot has grown massively on the strength of the company's niche: It consolidates quantities of products from about 1,000 manufacturers, purchasing them in less-than-truckload orders—ones that the manufacturers would otherwise not fulfill because they wouldn't be economical—and warehouses the goods. It then allows its customers to purchase as much or as little as they want of any combination of products, saving them from carrying large inventories themselves that they are not able to afford. One benefit of this model is that small brands that can't afford to sell to customers outside their region get more products to more people. That's because Dot Foods operates thirteen distribution centers, including two in Canada, in addition to its headquarters and warehouse hub in Mount Sterling, Illinois, boasting 4.2 million square feet of warehouse space in all. Dot Foods is the Amazon of this business, trucking custom on-demand orders faster than the competition due to staying ahead of the logistics innovation curve.

The result? Dot Foods has annual revenues of about $8.5 billion with a workforce of 6,400 employees.

Eight in ten US businesses are family owned, and those companies account for more than half the American economy's gross domestic product while also employing roughly six in ten workers. Some have grown into behemoths, such as Walmart, Cargill, Publix, and Ideal Industries. While most don't make it to a second generation of leadership, either folding, going public and taking on outside leadership, or being sold to other firms, those whose leadership positions get passed on to family members are an especially impressive lot.

Many of the exceptionally well-performing companies in the Titanium Economy are family run, and they're a far cry from the proverbial mom-and-pop corner stores and local restaurants that the term "family business" tends to call to mind. As we've seen, many are sprawling businesses that compete with some of the largest companies in their sectors while others have specialized niche segments, and many have sustained their success for multiple generations. We've found in our research that these family-run enterprises have flourished in large part because they blend a strong sense of purpose and culture with a long-term focus on customer-centric value creation as the driver of growth. In short, they don't suffer from "short-termism."

We've found that these firms demonstrate a striking commitment to building high-quality cultures that nurture employee skills and boost engagement. A raft of new research has found that these enterprises, more than other businesses, instill a feeling of "belonging" among their employees—a sense of inclusiveness, security, and support. This plays a critical role in workplace cohesion. Increased team cohesion in the workplace has been shown to boost success, work satisfaction, and self-esteem among team members while decreasing anxiety.

This feeling of being valued and connected also fosters innovation. The focus on longer-term growth supports making the considerable investments in new technologies that enable them to evolve. These companies are oftentimes more adept at adopting advanced industrial technology, like robotics and AI, than large publicly owned concerns. They're in it for the long haul and it shows. Dot Foods exemplifies these traits.

Wearing his customary black Dot Foods polo shirt, CEO Joe Tracy told us about how Dot Foods' competitive advantage wouldn't amount to much without fostering a pro-worker culture and endeavoring to understand the job experience from the employee's perspective. "Eighty percent of our employees are blue-collar warehouse workers and drivers," Tracy said. "The pandemic is by far the most challenging labor environment I've ever seen. Our volume is off the charts, and we can't staff it because we're typically one hundred drivers short and probably 500 employees short."

Staffing shortages, while exacerbated by the Covid-19 pandemic, have long been a problem in the business—something Tracy is all too aware of—and he has been nimble in finding ways to fill the need. "Workers want flexibility, so we've gotten creative with shifts," he explained. "Instead of five eight-hour days, we have four ten-hour days and staff three twelve-hour shifts on the weekends." While the schedule can be grueling, it allows for people who prefer extra days off. Some of the company's warehouse employees created a Facebook page that allows workers to contact one another to swap shifts and manage schedules in much the same way Uber drivers and other gig-economy workers do. "If someone doesn't want to work tomorrow, they don't have to work tomorrow," Tracy said. "That works for our workforce and has helped us to keep pace with demand." He emphasizes that younger workers, in particular, "typically care more about accruing time off than making additional money."

Dot also leverages robotics and automation to compete. "No one wants to work in the freezer at night or on the weekend throwing cases around anymore," Tracy said. Younger workers, especially, just don't want to do certain jobs, he points out, and research shows they have been leaving jobs to pursue other opportunities faster than older generations of workers. "But they're good workers, and we want to keep them," Tracy continued. "So we invested in a system that moves those cases." As some research has stressed about automation, it will be enhancing the quality of work life for many, and that's absolutely true in the Titanium Economy, which has its fair share of physically demanding roles.

As for worker development, Dot invests considerably in training workers. Like Amazon and other mega-warehouse companies, Dot hires warehouse applicants without a resume, and no previous experience is needed. The company teaches them to be certified in forklift driving and other skills. Regarding innovation, Tracy said, "We have embraced technology more than most distribution businesses have because family businesses that rely on the models of the past to carry them into the future just don't survive. In my eyes, a business must reinvent itself every twenty years or so and come up with a different business model in order to keep pace, and that's what drives not just me but our whole team at every level." Dot Foods has worked hard to integrate logistics with advanced analytics so that customers receive the products they want as fast as possible. It's precisely that mentality that has allowed Dot Foods to expand its product lines and grow its customer relationships.

Tracy explained that facilitating better access to information and enhancing variety has become crucial to the distribution business in recent years. "Technology is changing distribution in ways that might not replace our core business," Tracy said. "But it has the potential to drive the gross profit and the net profit for the items that really carry the day. That's why we have made the transition in our business from being logistics experts that know something about technology to becoming really good at technology while also happening to be logistics folks." The core business of loading a truck and shipping the product somewhere won't change; however, the right technology can optimize the route path for trucks to travel fewer miles here and there, adding up to reduced costs for fuel, labor, maintenance, and insurance.

As a private company, Dot Foods keeps governance information close to its vest. But Tracy and his team hired a chief information officer and have ramped up tech-related hiring in recent years. Despite being a transportation guy at heart, Tracy is confident that the key to the future of his business rests with programmers who can bring their large offering online, allowing customers to easily scan through the enormous inventory the company offers. "If that's what the business needs to thrive, that's who we'll hire," he said.

Such talent created Dot on Demand, which now makes the more than 43,000 in-stock warehouse products available to its customers via one platform, facilitating distributors getting in-demand products onto shelves fast.

But Dot's assistance doesn't stop at merely supplying products. In June 2021, Dot Foods acquired ShopHero, a grocery-focused e-commerce technology company based in Utah. ShopHero allows local grocers to quickly establish an online shopping experience with a turnkey solution: a customized, locally branded e-commerce platform that includes a web presence, a mobile phone app, and access to expanded online shopping options. "We have to have videos and images—not one image but five images. We have to get better if we're going to be able to help our partners compete in a digital environment," Tracy said.

"This is still a journey," he reflected. "Many family businesses in the industrial sector have to do things much differently than they have done in the past. That's high-risk, and there isn't a lot of room for error. It's not surprising many struggle and feel as though they could end up becoming irrelevant to their customers." Family business leaders also carry the added weight of honoring and burnishing their family legacy. For some inheritors of businesses, the passion for the entrepreneurial vision of the company's founder doesn't burn as strongly. But when next-generation leaders embrace the vision, we've found that they tend to feel an intensity of commitment to the values and purposes that drove the success of their forebears.

Vikings and Gardeners

Karen Norheim, CEO of American Crane & Equipment Corporation (ACECO), is just such a next-generation leader. Though she hadn't planned to take over her father's role as CEO, she developed a passion for manufacturing on her own time and in her own way.

Her father, Oddvar Norheim, emigrated from a small Norwegian island in the mid-twentieth century while in his twenties and found a job at American Crane. "He came from nothing," Karen Norheim related, "but he ended up taking the company over. It was a fledgling company that wasn't doing very well, and he turned it into a thriving enterprise."

ACECO specializes in manufacturing custom-made cranes, hoists, and lifting equipment for specialty jobs, which has made the company ubiquitous in manufacturing circles. These aren't the run-of-the-mill machine contraptions you find on a city street lifting cement pallets. The company produces custom equipment for factories, secure government facilities, and other manufacturing plants.

While her father was busy running the company, Norheim attended Penn State University, studying marketing and international studies, and she later worked in the ski and real estate industries. Like many young people, she wanted to find her own way, until 2003 when Oddvar asked his daughter to join him at ACECO. "At first, I didn't think manufacturing was all that cool or interesting,'" she recalled with a laugh. "But I had very happy memories of visiting my dad on the shop floor and breathing in that smell of the machines working. And, of course, my dad was my role model. I worshipped him. So I said, 'Okay, I'll come out and give it a try, but if I don't like it, I am out.'"

Norheim started her tenure at ACECO in marketing and IT, honing her tech skills, and when her father decided to slow down, she was elevated to president and chief operating officer in 2019. She expresses total clarity about the legacies of quality production and nurturing a culture of appreciating the employees she inherited, and she is adamant about honoring and building on that cultural transformation. Her father decided to keep operations in the United States so that he could ensure quality control. He explored moving operations to China, but in traveling there, he found that he didn't feel he could manage and oversee the high-quality product he demanded from afar. His daughter echoes that same ethos today.

"We compete against companies that make billions of dollars and have thousands of employees," she said, "and we've managed to keep our little niche here by biting off toes." What Norheim recognizes with this statement is the importance of leading in your micro-vertical, where small players can be successful. "We've stayed strong by making sure the quality of our components is consistent. It's something we all feel pretty confident about."

Regarding the company culture, she said, "My father's philosophy from the beginning was always that our employees are the most crucial part of the company. I always tell people that we are Vikings, but gardeners too. Essentially, we're fighting the battles in business while also tending our garden at home. Good people are our number one, most important asset. And if you have good people, that allows you to go out and fight the battles of the business and be successful."

As is a common concern when the next generation takes the reins of family firms, the question of whether she would continue to cultivate that ethic loomed as she transitioned into her leadership role. "People wanted the assurance that I would be around," she recalled, "and that I'd lock in my dad's legacy of putting perseverance, heart, integrity, and a strong work ethic first." She realized that the company's written statement of its values and purpose was outdated, and she initiated the process to make it more robust.

Norheim believes that her job is to be the chief reminding officer. "What we do, we do well; we do it on schedule, and we always follow through," she said. "So I'm here to remind people why we are here, what our vision is, why we do this, and why it is important."

Not surprisingly, she's an ardent supporter and promoter of applying technology to remain competitive. In 2019, she created an in-house innovation lab, where employees from different areas of the company, such as engineering, sales, marketing, and service, could come together to brainstorm and experiment. While she gave them the mission, some "tools and toys," and the space to see what they could build, she said the crucial piece of the puzzle was that she "kept my sticky fingers out of it," making the innovation lab employee-led. The company launched several initiatives leveraging Industry 4.0 technologies, including IoT, AI, VR, and more, to stay competitive in the ever-changing digital landscape. "It's important in the digital industrial space to not try to be the expert, but to collaborate," Norheim said of the work her innovation lab team does. "I want to empower people to figure out the solution to a problem."

Like many leaders in the manufacturing sector, Norheim is acutely concerned about hiring and retention. As a result, she's created a talent pipeline internally by investing in her workforce and supporting cross-training programs so that her people have an array of skills to carry them into the future. "We don't just have people who can weld, we have people who can weld and fit," she explained. "We don't have people who are just electricians. They are electricians, of course, but they can assist with mechanical assembly too. And once our team members master multiple skills, we then try to bring in technology as a complement that helps them to do their jobs better, or at least to do more than they might normally be able to do."

The program strengthens job security for employees by augmenting their skills for the future and instills loyalty as a result. So even though Norheim predicts that ACECO's shop floor will have robots within two years, she's confident that her team doesn't fear the change. She knows they understand that automation will empower the company to keep competing and growing. "My team sees it as an opportunity, not as a threat, and our culture plays to that," Norheim said. "Because it is a culture of trust."

Bucking Short-Termism

The leaders of family firms have the authority to pursue not only earnings goals but ones that reflect the family's desires. "In a publicly held company, your responsibility is to think about your current shareholders," said Meghan Juday, former chairman of the board of family-run Ideal Industries. "But we can afford to think about our future shareholders." If you wonder whether these nonfinancial motivations hinder performance and innovation, we can assure you that they do not.

Juday comes from the fourth of five generations involved in running Ideal Industries, which manufactures tools and supplies for the electrical market and battery-charging capabilities and maintenance in the infrastructure markets. J. Walter Becker founded the company on a set of corporate values that the family has maintained, though not always easily, as we will see. "J. Walter Becker felt that a company was more than just the bottom line—that there had to be multiple bottom lines," Juday explained. "He really felt that relationships mattered and treating people well mattered. So he came up with this name, Ideal, which indicated the sort of relationships he sought to have with his employees, customers, communities, and suppliers. If they all were treated well, he believed, then our company almost couldn't help but be successful. And here we are, 106 years later, still going strong."

Ideal is 100 percent family owned, with fifty family members accounting for thirty shareholders, many in roles on the Ideal Foundation, corporate board, or in family governance. While no family members are currently working in the business, they take their responsibility of company stewardship seriously. As you can imagine, decision-making can get complicated, and Juday was instrumental in finding a solution. She started working for the company in 2003 as a consultant. "I was taking leave to stay home with our newborn when my father called me and asked me to work on the transition from the third generation to the fourth as he was retiring," she recalled. Despite the burden of caring for a three-week-old, she knew how important Ideal was to her father and their family, so she agreed.

Her father put her to work cultivating the engagement of her cousins and others in her generation. "We were all in our twenties and thirties at the time," she said, with varying interest in the company. She also had to attend to the problem of conflict among some family members. To address that, Juday came up with the idea of reshaping the family council to address concerns, educate, and engage the entire family. "We had such conflict and so many challenges with a few family members," she recalled, "that I was certain we had about a year or two before we were going to start seeing lawsuits or massive sellouts." Instead, the fourth generation took over family leadership, and later board leadership too, which gave the company a firm bedrock from which to move ahead. "My father hasn't been chairman for eight years, and we are still managing the transition for his various roles and responsibilities," Juday said. "It takes time to allow the company to grow and absorb these changes as they evolve." The longer-term thinking of many family companies has helped the Ideal management team provide the latitude for that adaptation. Family businesses often run into challenges unique to their internal dynamics and histories, such as informal cultures and structures that don't support a robust business, lack of succession planning, lack of training, and pressure to hire family members, not to mention conflict that rises above business issues. About 40 percent of US family-owned businesses turn into second-generation businesses. The odds get smaller after that. Thirteen percent remain viable in the third generation, and 3 percent stay viable in a fourth generation or beyond. But they do have the distinct advantage that they can, as pointed out by Juday, whose family business has defied the odds, "see things in generations—not just quarters."

Most significantly, perhaps, decisions don't need to be passed up an intricate chain of command. Therefore, decision-making can be sped up significantly in the face of any disruptions, large or small. And while a diverse, experienced, hands-on board is crucial, when necessary, family-run industrials can make decisions on a more informal basis—something that is crucial to twenty-first-century Titanium Economy management since these companies can be more agile when they don't have to navigate bureaucratic corporate structures to make decisions on the direction of the business. These decisions can happen at the dinner table or the weekend family get-together.

Juday emphasizes Ideal's ability to privilege values-based decisions about how to grow the business and invest in the experimentation that is so vital to agility in innovation. "If you're invested in a publicly held company, the return horizon is usually less than two years on any investment," she explained. "So if a company makes an acquisition, they have to pay it off in eighteen months." While that means investors may get a more immediate financial return when putting their money into a publicly held company, they're often forgoing superior longer-term returns that result from firms making long-term investments. The financial benefits of those investments in the business may take at least three to four years to manifest, Juday highlights, and while that isn't acceptable to most investors in public companies, we suggest that they challenge their own short-termism. Like so many of the firms we studied, Ideal Industries has achieved phenomenal growth and returns.

As is also true with so many of these family-run firms, Juday and her family members have assured that their employees also benefit handsomely from these results. Providing high-quality jobs in eighty-nine cities and towns across the United States, UK, China, Canada, and New Zealand, Ideal has offered so many other families a stake in the long-term growth of the firm. The median income of an Ideal Industries employee is $43,000, compared to $31,000 in the service economy, and the company matches employees' 401(k) contributions with supplemental pensions so an employee who spends their career at the company can retire with 80 percent of their final year's salary.

When it comes to taking good care of their people, while all companies these days talk a good game, we've found that family-owned industrials tend to be doing a standout job of it. The commitment to their people feels more personal. That's not to say that family businesses don't have to sometimes make hard decisions regarding workers, but in that event, they tend to bring a face to the choices. Moreover, family businesses have a set of shared traditions and values based on their history with brands that are rooted in family legacies and reputations. And most of all, when they create value, it's with Main Street, not just Wall Street, in mind.

Family businesses like Dot Foods, ACECO, and Ideal constitute the heart of the Titanium Economy. Many have been around for more than one generation, and some have been in business even longer. The kids or grandkids are running the company now, and they're mindful that they're in charge of the family legacy. They want to do right. Whether they're now a global enterprise or are regionally focused, they're all about community. People know what their purpose is, they're very happy with what they do, and they feel an incredible sense of ownership. The competitive advantages offered to employees and investors that we've profiled through the stories of these three firms, which are representative rather than exceptional, boil down to four main factors.

First, the culture and social networking of family-owned businesses help employees motivate and protect one another and adjust quickly to new norms as needed. Second is the comparative speed of decision-making they are capable of, which is especially important in the face of disruptions that are all but inevitable for companies in every sector these days. Third, the regulatory and investor context for private companies, along with their freedom from investor pressure, allow these firms to be more agile and innovative in their strategic decision-making. Finally, family-owned businesses tend to carry less debt and have patient capital, or a long-term investment horizon. The companies have a long history, which lends a different perspective to the investment process. These enterprises and their principles are setting their sights on long-term value creation—and not generating high returns within a short period of time, which is a tendency of public and private equity models today. Oftentimes, these family-owned businesses are able to invest in projects that would seem "counterintuitive" to the market but result in the development of new innovations or performance improvements. During a downturn in the economy, they can continue to make investments and enter into acquisitions and partnerships at their own pace while private equity and publicly traded firms must heavily weigh investor demands. In a similar vein to the sector overall, this source of capital is undervalued and underappreciated and should be seen as one of the most obvious forms of available capital in the industrial sector. In addition to the patience and long-term nature of their capital investments, the family-owned businesses bring strong market and operating expertise honed over several decades, which can provide invaluable intellectual property to executives around the sector.

So the next time you note that an industrial company you're eyeing is family owned, take a closer look rather than dismiss it. These companies are where the hidden power of the Titanium Economy resides.

Chapter 8

Winning the Talent War

Early one morning in August 2021, a trio of women gathered outside a nondescript building in the City of Brotherly Love, ready to take part in a pre-apprenticeship program. One trainee had come straight from her overnight custodial job at the Pennsylvania Convention Center, where she made fourteen dollars an hour. When a friend told her about the program and encouraged her to apply, she replied, "Why not? I don't want to do housekeeping forever." Another woman was supporting herself and a ten-year-old daughter by waiting tables at a local diner, scrambling for tips to help make ends meet. She decided to apply to the program after asking for a toolbox for Christmas and realizing that she wanted to work with her hands. The third woman was an advertising sales representative for a local cable provider who had come looking for job security after having been laid off twice in four years.

Each of the trainees on that late summer morning had their sights set on the living wages, better benefits, and job security offered in the Titanium Economy, whose manufacturing workers earn 13 percent more on average than workers in comparable nondegree positions in the private sector. Not only do skilled trades consistently offer higher wages on average than the service sector, but industrial workers typically don't need a bachelor's degree, so to earn a good wage, they don't have to shell out $100,000 on average for four years of college tuition. That amount climbs to $172,000 on average for out-of-state schools. When you consider the loss of income during those years and the cost of student loan interest, the total cost of a bachelor's degree can exceed $400,000.

Yet despite the benefits, industrial employers can't fill open jobs fast enough, and the persistent talent gap for jobs at all levels has led to devastating results. In the past ten years alone, 2.4 million industrial jobs went unfilled, which was one of the primary reasons seven in ten companies fell behind in scheduled production. We estimate the US economy lost $2.5 trillion as a result of labor shortages. What's more, every industrial leader we spoke to for this book identified labor shortage as an impediment to future growth.

The six-week pre-apprentice class those three women signed up for is the first of its kind in Philadelphia, a trade readiness program offering a route to industrial jobs for underrepresented groups. Such programs have created a solid new pipeline of employees for many Titanium Economy companies, helping ease the talent gap. Through training and certification, they connect students with mentors, union representatives, and employers.

Getting more women into the trades might seem an easy path to filling the labor shortage, but those three women were outliers in their interest in the sector. While women make up nearly half the workforce, only 5 percent of female workers have jobs in skilled trades. Among electricians, that number falls to 3 percent, and in construction, it's even lower at 2.6 percent.

Women may not consider manufacturing jobs because they have so few female role models to introduce them to the fields. Industrial jobs have historically been dominated by men, so most young women don't consider careers in manufacturing. As well, both women and men may not realize how much the nature of manufacturing work has been changing for the better.

Karen Norheim, the CEO of American Crane and Equipment Corporation (ACECO) whom we met earlier, has found that a woman being the face of the company has helped greatly in recruiting women. "Too often, young women aren't aware of the opportunities available in manufacturing," Norheim told us recently. "But there is significant overlap between what young women want in careers and manufacturing today." Still, Norheim knows that her industry isn't considered glamorous, and she recognizes that the kinds of employees she wants to attract to ACECO would rather work for a tech company when they graduate.

She's making a concerted effort to raise awareness of the important work the company does. "You attract them by storytelling," she explained. "To get the word out, we share the journey of what we do. That's why I stayed. I learned our story. And

I realized that we do cool things. I mean, how cool is it that our engineers came together to design something that no other tech can do: developing cranes that safely move extremely hazardous waste? Or that we've been called on to build classified moving equipment for the nation's most high-risk projects, like building the cranes for disposal of nuclear waste left over from the Manhattan Project, helping to save the environment in the process. That's incredibly exciting. But it's a story that maybe isn't obvious to prospective employees at first."

Back at ACECO, to attract women who are a good fit for her company, Norheim appears in person and on Zoom at just about every recruiting opportunity she and her staff can identify, stressing the roles for women in manufacturing and the need for women to get advanced training in STEM disciplines. "There is exciting work to be done and a path that still needs paving for future generations of women," she tells them.

In April 2021, Norheim became the new president of CMAA, the Crane Manufacturers Association of America, a role in which she has shone brightly. The Manufacturing Institute, an advocacy and education group based in Washington, D.C., praised her as "a leader in the manufacturing industry not only because of accomplishments in the workplace but also because of her accomplishments mentoring young people, especially young women."

A frequent speaker at schools and professional organizations, Norheim prides herself on being a valuable resource for STEM programs in her region. She founded the Eastern Pennsylvania Chapter of Women in Manufacturing and quickly developed it into one of the premier chapters of the organization, which not only encourages women to enter manufacturing but develops the talents of women already in the industry through mentoring and peer-to-peer networking. That's important because getting more women into the industry is only part of the battle. Retention and advancement are also critical. In a field with a reputation as an "old boy's network," active mentorship by both men and women in positions of power is critical.

That's what Kim Ryan received. Ryan is the CEO of Hillenbrand, a diversified conglomerate making a variety of industrial materials with $2.9 billion in revenue in 2021. Early in her career, she learned the value of being mentored. "My boss would sometimes invite me to go to lunch at McDonald's," she recalled, "and that meant 'I've got another crappy job for you.'" When she confronted him about giving her assignments no one else wanted, he explained, "If I teach you how to fix things, if I teach you how to go in and learn what's wrong, work with teams on how to fix problems, and help those teams see that you can make them better, then you're always going to have a job. That's because every company has these pockets of problems, and every company needs people who are willing to tackle them."

She said these assignments helped her understand the company's business better and feel more attached to the company. "Projects and sponsorships cannot be underestimated when you're talking about how to get employees engaged for the long term," she explained. Highlighting a particular issue for women coming into an industrial organization, she said that if they are "shoehorned into a desk job where they don't have exposure to the company, there's definitely a high risk of a lack of attachment for those people."

In evaluating gender diversity in the manufacturing sector in 2020, McKinsey graded a mere 4 percent of companies as "diversity leaders." That meant they failed to either have 20 percent or greater gender representation in 2014 and improve on that number by 10 percent or more since then, or that they failed to have 30 percent or more gender representation in 2014 and improve the number by 5 percent or more. Some good news was that McKinsey also scored one-third of companies as "fast movers" in the hiring of women. These were companies with between zero and 20 percent representation of women in 2014 that had improved by 10 percent or more. That's a hopeful sign of progress, but much more must be done, not only in recruiting women, but in reaching young people across the board to enlighten them about and prepare them for work in the sector.

When it comes to ethnic diversity, McKinsey found that 9 percent are diversity leaders. That meant they either had 25 percent or greater ethnic representation in 2014 and improved on that number by 25 percent or more, or they had 50 percent or more ethnic representation in 2014 and improved by between zero and 25 percent. We scored about 30 percent of companies to be "fast movers," meaning they had between zero and 25 percent representation in 2014 and improved by 25 percent or more.

Dwight Gibson, the CEO of BlueLinx, agreed that the ways in which Norheim is attracting women to the industrial sector are crucial, and long overdue, and that they are also the sort of things that will attract more candidates of color to industrials.

As one of the very few Black CEOs of publicly traded companies in the United States, and among the youngest of that select bunch—not to mention the only one in the state of Georgia, as well—Gibson has experienced firsthand the same sort of evolution in the industry that Norheim has seen toward embracing female candidates for skilled roles at every level and helping elevate persons of color in the industrial sector. Still, he said, the sector remains overwhelmingly male and white and needs to do better if it wants to attract the talent it needs to turn good producers into superstars.

"I walk into a room, and half the time I can tell they're expecting someone else," he told us of his experience. "I was at a meeting with a vendor recently and someone asked me, 'So what do you do for BlueLinx? Are you the new marketing guy?' That still happens. But we're at the point now where I see that as their issue, not mine." He paused before saying, "It's been a challenge."

With experiences like that in mind, Gibson helped start the Black Employee Network at his previous company, Ingersoll Rand. Gibson's path to BlueLinx, a wholesale distributor of building and industrial products based in Marietta, Georgia, was circuitous, to say the least. The son of a single mother who was a teacher in Jamaica when he was born, his family ended up in the Bahamas before moving to Chicago.

"I'd never seen snow before," Gibson recounted with a chuckle. "I'd never had a winter coat before."

Finally, his family settled in what Gibson described as a working-class part of Brooklyn. "It was a good environment," he told us. "It was a mix of blue collar and white collar, but all very hardworking people. I made some good friends, positive kids, who really helped me make the transition. I learned a lot from that experience." He then landed at Howard, the HBCU, which he recalls as "incredibly diverse in terms of the diaspora" before a job here at McKinsey, where he almost immediately got involved in recruiting. He then had a stint at InfoUSA in Omaha, Nebraska, which was an early version of a data mining and supply company where Gibson spearheaded a move to the then-fledgling internet of the late '90s.

Gibson went on to work with BlackVoices, a Tribune asset, before heading off to Stanford and earning his MBA while continuing to support BlackVoices for a spell. After Stanford, he returned to McKinsey, where his last engagement provided an entryway to industrials. He then joined Ingersoll Rand before ending up running one of the companies that had been picked up through M&A, "just to see if I could do it." In short order, he'd turned the company around and realized, "This is what I like to do."

He ended up at Thermo King in Belgium, running the company's nearly half-billion-dollar interests in Europe, the Middle East, Africa, and Russia before heading back to the States and the company's Charlotte, North Carolina, headquarters. Throughout this time, particularly when Gibson was based in Europe, he was the most senior Black person at Ingersoll Rand, then a company of about 60,000. His next post was at the company SPX Flow before ending up as CEO at BlueLinx.

BlueLinx supplies more than 50,000 branded and private-label products within the building industry through over 2,200 associates and distributes a comprehensive range of structural and specialty products to 15,000 customers. In the past five years, the company has outperformed all of FAANG in terms of total shareholder returns, growing at 41 percent CAGR.

Under Gibson's leadership, the staff at BlueLinx has come to represent a diverse set of demographic backgrounds. As Gibson was quick to point out, the company's workforce is 30.2 percent female and 40.1 percent ethnic minorities and has great employee retention, with the typical employee staying on for six years and making a very competitive wage both for the region and the sector.

The way Gibson tackled the lack of familiar faces in the companies he'd worked for was, he told us, the same way he faced all the challenges throughout his nascent career. "I thought about everything in six-month or one-year increments," he said. "In my head, I was always thinking, *I'm going to get evaluated in six months. I want to make sure I'm learning as much as I can and adding as much value as I can.*" It's an ethos he tries to pass on to the remarkably diverse team he's built at BlueLinx. But he also insisted that the industry can and must do better.

"The first thing you have to do is help your employees, even recruits, understand the possibility," he insisted. "Not a guarantee; a possibility. And how do you do that? They need to see people that are different in the various roles within your company, whether it be Black people, women, or other minorities. You've got to make them really visible.

"Next, you have to help them understand how their capabilities and skills create value in a broader context, helping people to think more expansively about what's possible.

"Finally, you've got to go out and get talent from different places," he continued. "I've been in a lot of industrial companies, and I was always the first person to say, 'Hey, why don't we recruit at this mid-market school? Why don't we strike up a relationship with the local trade school? Why don't we go to an HBCU and try to attract talent?' Think of it this way: Does anyone ever ask why we're challenging assumptions or fostering creativity or collaboration? Because that's what diversity is. It's the soil in which all those things emerge. If you have different perspectives and people and backgrounds, you're going to get more creativity."

As for how he then helps those talented people find their way to the industrial sector, like Norheim, Gibson believes it's about the storytelling. "Again, you have to let people know what's possible," he told us. "That's what did it for me. If I could go back in time, I'd have that conversation of 'Dwight, I think you could do this.' Because, to be honest, I wasn't thinking about industrials. You've got to expose talented, diverse people to all the opportunities that are available. Companies have to be intentional about that. Leadership has to own that. Because the thing I'm most proud of in my career is the people I've promoted who have gone on to do really great things."

Young People Are the Future

In discussing the problem of the talent gap with us, Brady Corporation CEO Michael Nauman emphasized, "Manufacturing, as a subset of the economy, is totally misunderstood." He stressed that many millennial-age workers, as well as those from Gen Z, have never been in a factory, and they have little if any understanding about the nature of manufacturing work. What's more, as he highlighted, the teaching of trades in high schools has fallen off. He recalled that when he was in school at Brighton High in Rochester, New York, which is one of the top-ranked public high schools in the nation, "I took metal shop and wood shop." Many schools have discontinued those classes, and as for those that are available, he underscored, "In today's high schools, you must be literally seen as a derelict to get into any type of trades program."

Nauman, therefore, sees it as the responsibility of industrial leaders to impress upon youth that manufacturing companies offer them good career paths. But the US education system must also move beyond its one-size-fits-all approach to secondary school, he said, and offer more trade skills–based pathways for students, increasing enrollment in programs to train students in technical skills that are in demand by employers like Brady.

As for college, while we, of course, embrace the value of humanities, history, social sciences, and the arts, the United States needs more graduates in the STEM fields—science, technology, engineering, and math. Better options for post-secondary education are also needed. Too much emphasis is placed on getting a four-year university or college degree, which isn't for all high school graduates and has left many college graduates burdened by excessive debt.

Whether someone has a degree or not, there's a place for them in the Titanium Economy. If anything, skills rule the day and not college degrees. "White-collar and blue-collar workers achieve success together," as Brett Cope, the CEO of Powell Industries, put it. At Powell Industries, a manufacturer of integrated solutions and electrical equipment for the distribution of electrical power in commercial and industrial markets, Houston's diverse population offers a rich pool of recruits. After making a significant investment in recruiting talent, "we make sure our skilled laborers in the factory have every chance to succeed on a project," Cope said. "Whether they went to trade school or otherwise, the median salary across the company is $53,000."

Many of the leaders of industrial-tech companies shared this ethos, such as Meghan Juday of Ideal Industries. "We need a new dialogue around what it takes for a high school person to consider manufacturing, or to consider the trades, as a meaningful career," she said. "We need to bring back trade schools and stop talking as if the only meaningful way you can get a career and make money is to go to college. You should go to college because it supports your career goals. I'm not saying don't do that. But that shouldn't seem like the only journey for your average high schooler."

Many American parents agree. In fact, an April 2021 Gallup survey found that 45 percent of parents of current students wish that better alternatives to bachelor's or two-year degrees were available to their children. More than half the survey's respondents were interested in the prospects of apprenticeship programs, but they also responded that they were far less likely to say they know "a lot" about apprenticeships (9 percent) or technical training (22 percent), as opposed to those who said they knew "a lot" about two-year colleges (47 percent).

Making matters more urgent, the United States is being left far behind in schooling for industrial work by other competitor nations. Germany, Sweden, Norway, Finland, France, and Denmark all offer free or low-cost post-secondary education options tailored to work in the manufacturing sector. Germany has a particularly impressive system of Meisterschule, meaning Master Schools, that offer training in the trades and have partnered with industry to develop their curricula, with industry veterans serving as school principals. The country's "dual training" model provides a combination of academic classes in vocational schools, for two or three days a week, with on-the-job training in companies, preparing students for over 300 nationally recognized occupations. Fully half of those who finish school undergo vocational training provided by companies, which consider dual training to be the best way to acquire skilled staff. Half a million Germans enter the workforce through apprenticeship programs. Competition for them is fierce, in contrast with US perception, and being chosen for an apprenticeship carries the same prestige as higher university degree programs in Germany.

In Asia, Singapore boasts one of the world's highest-achieving educational systems. Multiple secondary school pathways place students in environments tailored to their abilities and direct them to post-academic employment opportunities. The field of polytechnics is one of the paths that high school students can take to develop employable skills. Due to these skills, a mere 4 percent of fifteen to twenty-nine-year-old Singaporeans are not in some type of employment, education, or training program, compared to 13 percent in the United States.

South Korea has built on the success of the German model to increase its emphasis on apprentice-based education and training. In 2010, Korea established its first Meister schools, partnering vocational schools with companies to tailor curriculums and courses according to the demands of the workforce and hiring school principals with real world business experience. The Meister curriculum areas of emphasis include biotechnology, semiconductors, automobile manufacturing, robotics, telecommunications, energy, shipbuilding, and marine industries. One of twenty-one Meister high schools in Korea, Gumi Electronic Technical High School demonstrates the strong bond with industry. With an executive member of LG Electronics as the school principal, Gumi has signed an agreement with LG Innotek that the company will employ one hundred graduates, and similar arrangements are in the works with LG Electronics and LG Display.

The United States must create more opportunities for affordable, or even free, post-secondary education, in addition to vocational training and undergraduate and master's education in STEM. We also favor career and technical education (CTE) in high schools that prepares students for jobs in specific industries. With a skills-focused training program that students can complete in one to two years, a high school graduate could step into a well-paying job immediately upon graduation. An added benefit of CTE programs, though, is that a substantial portion of students who graduate from them go on to attend two-year community colleges to further hone their skills.

Many business leaders and elected officials have now placed creating a "worker pipeline" for the industrial sector at the top of their agendas, and we have a great opportunity to accelerate renewed interest in CTE. But to attract the necessary students, we have to eliminate the stigma attached to this schooling, which can come from parents, teachers, and school counselors. Students need nonjudgmental information about the benefits of CTE. School systems should also hold district-wide awareness sessions with parents to show them how their children can better compete in the emerging Titanium Economy in decades to come. In addition to this counseling and outreach to parents, the United States must enhance the apprenticeship and training programs available.

For example, we should build on the work of the highly effective Registered Apprenticeship College Consortium. It partners employers with more than 300 colleges to align college offerings with workplace needs. The American Diesel Training Centers has pioneered another great approach. The training offered by the Institute requires only 20 percent of the time and cost of the typical technical school thanks to its pay-for-success model in which the cost of training programs is partly based on whether a student gets hired after completion.

Another promising effort is Generation USA, an independent nonprofit founded by McKinsey that offers free job training and placement services in such hot areas as cybersecurity and web development. In October 2020, Verizon committed over $44 million to the organization for a multiyear strategic partnership intended to significantly increase enrollment in the programs and enhance job placement.

Important as such initiatives are, we believe the country must make a concerted national effort to increase apprenticeship opportunities. We strongly support a nationwide "Apprenticeship Institute" network, which would elevate the importance of apprenticeships in the culture to the level of university degree programs, as in Germany. We hope that support for this approach would come from both sides of the political aisle. There is an opportunity for government and industry to come together to combat the perception that apprenticeship programs are inferior to university degrees, and as part of this, industries and government could fund a national campaign to rebrand apprenticeships to bring in new recruits.

The Creativity of Companies Making a Difference

Many of the Titanium Economy companies are providing great models for national efforts, taking matters directly into their own hands with an array of creative solutions. Remember the Casella brothers? Waste management may not be the first career choice for many people, but Casella Waste Systems has found creative ways to attract and keep good talent. For example, in 2018, the company partnered with the Stafford Technical Center to offer training for a commercial driver's license. The course brings students to Casella's West Rutland, Vermont, facility for eighty-plus hours of training on the company's waste removal vehicles while at the same time offering them a peek inside the doors to get a sense of its wider operations.

One of Casella's instructors, Bill Baptie, said, "Most of our managers and general managers started as drivers. So we're creating what could be a long, significant career for some folks. They get that license, and we'll pay for the course if attendees give us one year of service after they graduate."

The company also gives employees at every level a clear pathway to advancement. For example, a rear-load driver, hardly the most desirable position in the trash pickup hierarchy, can train to become a front-load driver and later advance to rollout driver, which involves dealing with large containers at job sites. "They can go from eighteen dollars an hour to thirty-five dollars an hour in about five years," John Casella reported, and he said they are building career paths for every position in the company.

The fluid-handling company Graco has also been proactive, building relationships with local trade schools and community colleges in Minnesota, South Dakota, Wisconsin, Michigan, Ohio, and Pennsylvania, where much of their critical manufacturing is based. The company has donated capital equipment to these institutions and offers scholarships. Graduates who work for Graco visit the schools regularly to talk about careers in manufacturing. Graco has also cultivated good relationships with professors in these schools, so the company's recruiters know who the top graduates are and can focus on recruiting them. Graco also has a successful internship program that helps it hire top performers. In addition, the Graco Foundation provides grants, employee incentives, and scholarships that support a variety of nonprofits, community colleges, universities, and high school apprenticeship programs. In 2021, that funding exceeded $1.3 million.

Another approach to making jobs in the Titanium Economy more attractive is the employee stock ownership plan (ESOP), which is funded through profit sharing and allows employees to earn an ownership stake in the company. CSW Industrials CEO Joe Armes shared with us how well the plan has worked for the firm. As with all Titanium Economy companies, CSW is up against steep competition from big tech for talent. "Obviously the Googles and the Microsofts have a significant leg up over us," he admitted. Bringing up the elephant in the room, he said he often jokes with workers, saying, "Well, we're in the grease business. I want to show you our grease plant out in Rockwall, Texas. It'll be the nicest grease plant you've ever been to."

But he's been able to attract great recruits with the company's ESOP plan, which he called "our secret sauce." He believes sharing wealth is the right thing to do and is good for business because when employees are owners, they have a more nuanced understanding of management's toughest decisions.

"We're not the kind of company that's offering ping pong tables, bring your dog to work, craft beers at lunch, or any of those types of things," he said with a laugh. "But our employees own about 5 percent of our company. We're all owners. Everybody participates in our success. We do not apologize about our quest for success and profitability, and when we succeed, we all succeed together. That is an incredible alignment of interests and a lever that we are able to use to drive value."

This ethic was demonstrated during the Covid-19 pandemic. "At the beginning of the pandemic," Armes told us, "we announced that we were not going to have any layoffs. We said,

'We don't have to.' We were plenty profitable at the time, we had no debt, we had cash on our balance sheet. We're not going to make a nickel or two more per share for two quarters on the backs of our workers."

As Armes has found, when employees have an ownership stake, they think of the business differently and take more pride in their contribution to success. Armes said he wants employees to feel like this is the best place they've ever worked and to tell their friends and family, "You ought to come work for these guys because they'll take good care of you."

Some companies are also providing training of their own, such as CaptiveAire. CEO Robert Luddy said, "If someone's willing, we can train them to very high levels." The company offers starter courses and then pairs employees with skilled workers to apprentice them. "If you take a young person and in ninety days you teach them basic vocational skills," Luddy explained, "then move them into an environment where someone's going to continue to apprentice them in those skills, soon you have a productive worker." He said this is preferable to going to community college for two years and then doing an internship. "It's too long."

Vanessa Tisci—who serves as group general counsel and compliance officer at Cavotec, a forty-year-old company in the clean-tech space—firmly believes that employees need to see the entire organization, including herself. A lawyer by trade and a curious person by nature, Tisci took a job with the retail pharmacy company Walgreens Boots Alliance and spent her first week working in a Boots retail location in the UK. "I wanted to understand the layout of the stores, how you launch a new product, how you market a new product. My legal team thought it was strange because they had never seen anything like it. They didn't understand why I wanted to do it. But actually, it helped me a lot with understanding the business and how contracts should be drafted in order to meet the business needs."

She also was mentored by one of the managing directors and quickly learned the ins and outs of the retail business world. "You can be a great lawyer in private practice, but you could be a total failure as an in-house counsel if you don't talk the language of the business," she explained. "So he started inviting me to all the business meetings. I realized that in-house legal work is not like giving advice in private practice. In the end, the essence of in-house legal work is to make the business grow."

She soon landed at Cavotec, her first industrial experience where she learned some industrial basics: "You need to control the supply chain. You need to control your margins. You need to control the cost of products, logistics, delivery."

To help grow talent and encourage a diverse pool of candidates, Tisci suggested implementing departmental rotations for new recruits. "Big companies have all these satellites. There is finance, HR, legal, procurement. They don't always talk to each other. And a lot of the time, they don't even know how to talk to each other because they don't understand each other. But how can I be a good general counsel if I don't understand the other parts of the business?"

That gap, she said, could be solved by such a rotation program. "When employees gain that sort of 360-degree knowledge, it will change everything about not just how they work together, but the types of people that are attracted to the company." Moreover, she insisted that millennial and Gen Z candidates will be attracted by the tangible results they will see when working in the sector. But, she insisted, the industry needs to do a better job at telling that story. "If someone has a passion for concrete things and seeing things get done, this is the right sector," Tisci told us while acknowledging that it's an underappreciated attraction of the industrial field. "It's not just about a software. It's not just about an algorithm. It's not just about data. You build something. That thing then gets sold in the market. That's an incredibly exciting story that isn't being told."

Bold Action Is Needed

The Titanium Economy's growth will stall without cracking the talent problem. To do this, the United States must think outside the box. A particularly creative approach to burnishing the image of industrial work, and attracting talent as well, is the Ideal Industries Electrician's National Championship, which offers competitors the opportunity to win some of $600,000 in total prize money. In 2021, to earn one of the handful of spots in the final round of the competition, entrants had to beat out 58,000 others who participated in qualifying events.

When electrician-in-training Mike Zurenda stepped onto the competition's stage in Nashville that December, wearing a bright-blue hard hat, ESPN cameras were rolling. In a series of challenges, competitors had shown off their skills as electricians, attaching wires and bending conduits with practiced hands, with the cameras capturing every move close-up. Professional announcers gave viewers a play-by-play worthy of, depending on your poison, Monday Night Football or the Great British Baking Show.

After four days of such challenges, Zurenda took first place in the Student/Apprentice category. When interviewed by his local television station, the upstate New York native said, "Hooking up some electrical device is the easiest part. It's getting the wires there; it's getting all of the materials set; it's getting everything looking good. Having the pride in your work to make sure you can stand back and say, 'I would want that in my house; I would want that in my building' would be the biggest thing to it."

He beautifully expressed the pride in craftsmanship that Ideal hopes to showcase to the world through the competition. Ideal's Meghan Juday calls it a way to let the world see what her employees and customers do every day, and she hopes it's the kind of exposure that can attract more talented workers like Mike Zurenda.

In 2009, the McKinsey Global Institute published a massive analysis on labor disparities that concluded 71 percent of US workers are in jobs for which there is weak employer demand, an oversupply of eligible workers, or both. The study concluded that unless American workers can develop new skills, the nation risks another period in which growth resumes but income dispersion persists, with Americans in the bottom- and middle-earning clusters never really benefiting from the recovery.

With the current wave of automation and AI disrupting production functions through better analytics and increased human-machine collaboration, the need for physical and manual skills is decreasing at more than twice the rate of that for the whole economy. In addition, the need for basic office skills is also declining as support functions are automated. Meanwhile, the number of professionals such as sales representatives, engineers, managers, and executives is expected to grow. This will lead to a greater need for social and emotional skills, especially advanced communication and negotiation, leadership, management, and adaptability. The need for technological skills, both advanced IT skills and basic digital skills, will increase as more technology professionals are required. Demand for higher cognitive skills will grow, driven by the need for greater creativity and complex information processing.

While we've made progress, the stakes have changed. Many CEOs and analysts agree. Through our industrial sector interviews, we have found unanimous support for dramatic and swift action to shift our education resources and policies to prepare Americans with the skills needed in the Titanium Economy. One key issue that is upping the urgency is the current crop of skilled workers nearing retirement. Juday stressed to us, "When you look at the average age of an electrician, it's fifty-five. The average age of a plumber is fifty-seven. So when you think about all that talent and expertise that is going to be retiring in the next ten years, you also have to remember that there is no way these new individuals who are currently coming up through the trade schools will be able to replace that talent."

The country must seize the opportunity to reorient perceptions of Titanium Economy jobs created by the growing consensus among business and government leaders about the need to do so. The Covid-19 pandemic has also presented an opening for changing minds, as it has forced many people to reevaluate their personal and professional priorities. By showcasing the rich opportunities for meaningful, reliable, and well-paying work in the Titanium Economy, companies may be able to attract many of these people who are interested in pursuing new life paths.

Fulfilling the potential for growth in the Titanium Economy hinges not only on attracting more young people to industrial careers but also on retraining the current workforce. With automation already ramping up and accelerating, the nature of the jobs many of them are doing will be evolving. We need to create a groundswell of interest in making the United States the envy of the industrial world once again by capitalizing on our greatest asset: putting more of America's talented workers into good-paying careers, one industrial at a time.

Chapter 9

Sustainability Should Be the Last Word

Hour by hour, trucks roll into the Trex factory in Winchester, Virginia, a 200,000-square-foot complex the size of four football fields that runs around the clock, seven days a week.

And it has to. Trex's contracts with local big box stores, such as Target and Walmart, and grocery stores like Albertsons, as well as drop-off recycling centers around the region, guarantee that the company will accept any plastic film waste sent its way. This includes all the grocery bags, dry cleaning bags, and newspaper sleeves, as well as the plastic wrapping around paper towels and any other plastic wrapping that consumers discard daily. All that would otherwise rot in landfills, becoming brittle, cracking, and creating microparticles that seep into waterways. Trex instead converts them into the raw materials used to make its highly durable alternative-wood decking through an inspired process pioneered by Roger Wittenberg, a petroleum engineer and one of Trex's cofounders.

The company received the first patent of its kind for the process in the late 1990s. Wittenberg's first creation, a park bench, proved the viability of the material and inspired him and three partners to found Trex in 1996 with the mission of manufacturing sustainable building materials on a large scale. Honed over the years, the process Trex now utilizes for its decking, as well as for railings and benches, combines plastic with reclaimed wood to produce planking that looks remarkably like wood but is more resistant to weathering and decay.

Trex's intake of plastic bags has grown so vast that the company's IT team was forced to refresh its computing capabilities in 2020 to keep up with inventory, not only at its Winchester facility but also at mega plants in Nevada and Arkansas. Either the company's onsite data center would need a complete hardware upgrade, or Trex could switch to cloud computing. Both would involve a substantial investment, but moving to the cloud was the forward-thinking option, so it was the clear choice for the innovative firm.

If your only experience utilizing the cloud is via Google docs or when accessing iPhone photos, imagine a thousand more specialized functions, from security to inventory management, driven by many different pieces of software, some custom and some off-the-shelf, with different access protocols for thousands of people, and you'll have some sense of how elaborate the upgrade was. The new cloud computing setup gives Trex all the capacity it needs to keep growing through an extensive network of servers that can store, manage, and process any amount of data and conduct whatever analytics the company throws at it.

Outdated IT infrastructures are holding many manufacturers back. They haven't been able to invest the considerable funds required to upgrade hardware equipment, and most factories also can't afford to shut down to allow that upgrading to be done, even briefly. Cloud computing facilitates quick upgrades while offering manufacturers immense computing power and more robust tools than they would otherwise have access to. With the cloud, Trex employees can access any internet interface, reducing the frequency of hardware upgrades. Tools are also faster, more dynamic, agile, and more user friendly, allowing for more data collection with insights that are easier to disseminate across the company. The massive upgrade undertaken by Trex was rolled out quickly. Best of all, Trex's cloud upgrade will allow it to easily scale up its computing capacity in the future for cranking up production or opening new factories.

All of us, across the planet, have a stake in Trex's growth because the company is doing so much to cut down on some of the most problematic types of plastic waste. The bags and wrapping materials it repurposes are generally not accepted by recycling programs because they gunk up sorting machinery and are not made of high-enough-grade plastic to make recycling them economical for most purposes. They make their way not only to landfills but into rivers, lakes, and oceans, where they endanger animals that have been found washed ashore with appalling amounts of plastic in their stomachs. What's more, the microplastics they release into waterways make their way into drinking water. This makes the remarkable ingenuity Trex has brought to the continual innovation of its processes a vital model for the Titanium Economy's potential to spearhead enormous progress in achieving sustainability goals.

Jay Scripter, the former vice president of operations at Trex, explained that the company invented a breakthrough method for utilizing "corrupted" plastic bags, meaning ones that have been soiled by food and other contaminants. "We created a process that could accommodate what was essentially garbage," he stated, "rather than pristine materials," which was key to making the use of waste plastic economically viable. The process meant that instead of having to sort dirty bags out from the masses of material Trex collects—from grocery store chains, universities and schools, and many other types of organizations—it could process almost everything. Once the bags are accepted, machines break apart and grind the plastic material into granules before mixing in sawdust from recycled wood shavings along with dye to produce an array of wood shades.

Every sixteen-foot plank that comes off a Trex production line contains approximately 2,250 plastic bags, and the wood shavings that are used come from the waste product of other manufacturers. This means that Trex hasn't been responsible for cutting down a single tree to make any of its products. Every year, the company salvages more than 500 million pounds of plastic and wood scrap that would otherwise have gone to landfills.

Trex has also pushed to make all its production processes as sustainable as possible. Its production methods eliminate smokestacks, and any leftover materials are recycled back into the manufacturing line. Scripter has stated the company has even sought to address noise pollution caused by its plants, installing noise mufflers on the roofs of nearby houses. The packaging material Trex uses for its products is also made from recyclable materials.

The Titanium Economy is alive with innovation seeking to reduce the environmental footprint of its products and the processes of making them. In fact, every company we examined, from old-line manufacturers to upstarts, is working on sustainability efforts, inventively using emerging technology to supercharge progress in many of the sectors that can make the biggest difference in remediating pollution and climate change.

The building and infrastructure sector, to which Trex belongs, has the potential to make the biggest impact among industrials in reducing greenhouse gas emissions. According to the World Green Building Council, construction and materials account for 11 percent of global carbon emissions, and a host of new methods are being developed for building offices, municipal complexes, and homes that are more environmentally friendly. These include the invention of new materials, such as a range of sustainable replacements for concrete, the manufacture of which alone accounts for an estimated 8 percent of global carbon emissions. New methods for recycling glass, which has been a challenge for the recycling business, are turning masses of bottles that would otherwise go to landfills into materials for infrastructure, such as roadways.

Another one of the sectors in which Titanium Economy innovators are contributing to great progress in sustainability is agriculture, which is also high on the list of the biggest greenhouse gas emitters.

A STRAIGHT LINE

Ask any farmer the secret to a good season, and they'll tell you the straighter the row, the more efficient the harvest. As one of our colleagues who grew up in Minnesota farm country told us, "Back in the day, being able to plant in straight rows gave you bragging rights. I remember my dad and his friends arguing about who planted the neatest rows, which was the measure of whether you were a good farmer. They took it very seriously."

These days, farmers can get nearly perfect rows with the push of a button. In one example, John Deere's fully automated 8RX Tractor comes with a sixty-foot planter that can seed twenty-four rows at a time. Farmers only need to drive this advanced piece of machinery once around the border of their fields, and it can then create a template for automated planting. During this pass around the field, 300 sensors combine with GPS-powered satellite technology to make up to 15,000 measurements per second in constructing the template. Then, at the push of a button, that same tractor will drive itself, plowing and seeding rows in the straightest lines that science can calculate and uniformly spacing seeds, heightening precision placement and increasing crop yields.

That sort of gathering and utilization of data is advancing a revolution in farming known as precision agriculture. It harnesses not only sensors and satellite surveillance but drones, artificial intelligence, and robotics, as well as data analytics, to make farming more efficient and healthier for the planet. Achieving substantial gains in yields is another key goal, made pressing by the expectation that global food demand will increase 50 percent by 2050 due to a steadily growing global population.

An explosion of technology is being developed to meet the challenge. Deere tractors come fitted with rearview cameras and touch screens showing fertilizer and pesticide application rates and placement, and they have better fuel economy. GPS-powered autonomous guidance ensures efficient coverage with accuracy down to sub-one-inch specifications. These self-driving tractors can plant even on the darkest of nights, ensuring that seeds get in the ground exactly when they need to, which can be a difficult challenge for farmers. The optimal planting window for most crops is just ten days, and with weather conditions often interfering with planting, getting all fields planted in time is often a race. The tractors also free up farmers' time for the many other tasks that make what they do such a demanding profession.

Precision equipment also crucially reduces the amount of water and environmentally damaging chemicals needed to operate today's farms. Preserving water supplies has become an urgent issue, with many of the aquifers that supply so much agricultural water drying up. Meanwhile, the use of high volumes of chemicals has dramatically depleted the health, and therefore the fertility, of soil in many places, with chemicals also draining into water tables and causing environmental scourges such as massive algal blooms.

John Deere has been a leader in the advent of precision agriculture. In addition to bringing smart technology to its equipment, the company is a leader in after-sales service, meaning ongoing services provided to customers after the purchase of products. Deere offers data services that generate customized feeds of information that are delivered directly to the home computers and tractor touch screens of their customers, allowing farmers to closely monitor crop growth, soil temperatures, and moisture levels. Deere also scaled up its dealership network with the aim of providing the best technical support in the industry through its Connected Support service. All this innovation has made Deere the industry's primary platform for equipment hardware and software.

Deere's reinvention of itself into a high-tech industrial has been remarkable, even for a company founded on innovation. The brand launched in 1837 in Moline, Illinois, which is still the company headquarters, selling one of the first commercial steel plows, which was invented by its eponymous founder, John Deere. But even for a company that has constantly innovated ever since, redirecting a now $37 billion, multinational behemoth was no small feat. The process of transformation exemplifies the longer-term thinking of so many Titanium Economy firms. For example, back in the 1990s, Deere bought a military-grade GPS system well before GPS technology was being broadly commercialized. Over time, it gradually and methodically integrated satellite guidance into its tractors, making sure along the way that its dealers and best customers understood and accepted the advancement. It also developed the tractor-mounted StarFire 6000 satellite receiver, which can be installed on tractors of other brands and is now standard equipment across the industry.

Deere isn't letting up on innovation. It recently opened its precision tech platform to third-party developers, as many Silicon Valley tech companies have done, and it has partnered with NASA's Jet Propulsion Laboratory on continuing to advance self-driving farm equipment. "Rather than laggards, they are leading tech and holding their own against the startup ankle-biters," said noted equity analyst Stanley Elliott, "and they are getting the VC money to fuel their ongoing drive for innovation. Look for tractors to be completely automated in the future."

Deere has also innovated in its sales operations, creating an online virtual product pavilion that allows prospective customers to, in the words of the company, "See what's under the hood. Step inside the cab. And even kick the tires," all in vivid 3-D imagery.

Deere has also worked hard to maintain strong relationships with its dealerships and ultimately customers as it has evolved so rapidly. The company has been a strong presence for generations across rural and suburban North America, with offices and dealerships having become common sights. Moreover, company sales representatives and other employees are Deere ambassadors in their communities, staying in tune with their farmer customers when they run into one another at high school football games, church pews, or the feed store.

Deere has continuously created jobs and fostered goodwill in rural areas across the United States. It's no wonder, then, that so many farmers will tell you that they "bleed green"—their way of referring to being loyal to the brand that's distinguished by the bright green paint of its machinery. Troy Taylor, the owner of one of the company's largest dealerships, Tellus Equipment Solutions of Katy, Texas, expressed this sentiment, saying, "John Deere represents the type of brand that me and my family wanted to be associated with. It's a long-standing, top-tier brand, known for its precision, its quality, and its customer service."

Still, like so many US companies, John Deere fell on hard times in the early 2000s. The pivot it made to digital services and new revenue streams was instrumental in how it bounced back. Deere has since won over its base of investors, tripling its stock price since 2017. Investors see potential for a tremendous additional upside in the long run while, in the near-term, commodity prices, weather, and other exogenous factors still make them wary. Agriculture is, after all, one of the most volatile of all industries. Here, too, Deere is making bold moves. "Most companies trying to grow profits focus on increasing revenue or lowering costs," Elliott observed. "Deere is doing both things at the same time for their farmer customers. The result is that they are less dependent on commodity prices."

Deploying advanced technology to improve farming yields in rural communities that have suffered great economic distress in recent decades is a mission that's vital to the United States and the global communities Deere serves. Its high profile as a technology leader is helping it attract the sort of talented workers who are typically more interested in settling on the coasts than in Moline, Illinois.

The Miracle Box

If solar panels covered just 22,000 square miles of the United States—an area smaller than the size of Lake Michigan—they would generate enough energy to power the entire country. Once the purview of commercial buildings and expensive homes, the price of solar panels has dropped dramatically in recent years, making them more affordable to average homeowners. Nevertheless, obstacles stand in the way of more people embracing the technology.

One issue is the Christmas tree problem in which a single dead light in a string takes out all the other bulbs. Similarly, shade or dirt obstructing one solar panel can hinder all the panels in a chain from absorbing energy since each row is aligned in a single series and all the panels are connected. A drop in voltage affecting one panel can cause the whole system to fail. Additionally, because roofs are subject to bad weather and debris, the system needs to be able to withstand lots of abuse.

Enter solar technology manufacturer Enphase Energy, which employs about 775 employees at manufacturing facilities in several states, including Texas, Idaho, and California, where the company is headquartered. Enphase sits in a sweet spot of the Titanium Economy, with a breakthrough product in one of the fastest-growing segments. The company invented a new kind of microinverter, the device within systems that converts the direct current that panels absorb from the sun into the alternating current we use in our homes. It solves the Christmas tree problem.

The Enphase microinverter is a tiny device attached to each panel, whereas converters have traditionally been larger devices that converted the energy for an entire set of interconnected panels and were prone to breakdowns, which disabled the whole system. With the Enphase technology, if one panel's converter is disabled, other panels are not affected. The Enphase microinverters are also more cost-effective.

The brilliance of the innovation was underscored by TJ Rodgers, a noted scientist, venture capitalist, and early investor in solar cells, who is also on the Enphase board. "It's hard enough to get a power plant to make that conversion," he explained. "To get a hundred-dollar microinverter to do the same task requires a high level of design." He pointed out that the task required designing customized computer chips. Part of the beauty of the accomplishment is that, as he also explained, "Because everything is connected digitally, you can keep watch on each inverter. You can, for example, examine the power output of every panel on your roof, and if one panel starts to fail, you'd know it."

Like so many Titanium Economy stalwarts, Enphase was built as much out of intuition as it was opportunity. Frustrated by the low performance of the string inverter for the solar array on his ranch, Martin Fornage teamed up with Raghu Belur, an engineer at Cerent Corporation, in the early 2000s to develop the new microinverter. By 2007, the pair had developed a prototype, and they teamed up with Paul Nahi to launch the company with Nahi as CEO. After tapping approximately $6 million in private equity in 2008, Enphase released the first iteration of its microinverter to moderate success. The second-generation microinverter, released in 2009, was far more successful, with sales of about 400,000 units as the decade waned, and Enphase gained a 13 percent foothold in the market supplying residential solar systems by mid-2010. The company's growth skyrocketed from there. Enphase shipped its 500,000th inverter in early 2011 and its one millionth later that year. Combined with the third-generation microinverter, released in the summer of 2011, Enphase sold over one million units that year, bringing the number of microinverters in use to 1.55 million for a 34.4 percent market share.

By 2012, Enphase's inverters had captured a 53.5 percent market share for US residential installations, which represents 72 percent of the entire world microinverter market, making the company the sixth largest inverter manufacturer worldwide. In turn, Enphase branched out into the burgeoning solar markets in Europe and found considerable success in the UK and, eventually, Australia.

Meanwhile, back here in the United States, Enphase continues to grow both its market share and its capabilities. In another breakthrough, Enphase has made its microinverter so energy efficient that it doesn't need a cooling fan, a considerable cause of expense for most arrays. And by continuing to improve the quality of its production, the microinverters have achieved "semiconductor-level reliability," according to Rodgers.

Enphase is now helping lead the way in making residential solar power pay dividends for customers. It installs sensors to networks of solar panels that enable customers to feed energy back to the grid, earning them income from their local utility. The next step for residential solar may be the creation of microgrids, which combine panels with battery storage to allow homes to go off-grid, such as during a power outage. Another advantage is that the battery capacity allows for using solar-generated electricity during the hours when the sun isn't shining, which has been a significant limitation of solar systems. Rodgers has transformed his own house into a microgrid. If his electricity provider suffers an outage, he can simply flip off the connection to the utility grid and turn on his home system. Numerous homes with such battery storage can be connected to form a community microgrid, with neighbors buying and selling power to one another.

This can solve the problem, Rodgers explained, of utilities not being able to "afford to buy acre after acre for batteries," which has meant that their large-scale solar fields have been limited to operating during daylight hours. With households buying their own batteries, the benefits of solar will become increasingly clear to the public. A big motivation behind people's investments in these battery systems is that consumers are demanding greater reliability and affordability.

Whatever fuels faster adoption of solar electricity is to be loudly applauded; the reduction of fossil fuel emissions needed to mitigate climate change can't come soon enough.

CREATING ENERGY FROM WATER

The world-changing potential of the green hydrogen fuel industry is making energy upstart Ohmium friends in high places. When the government of India went looking for ways to modernize its energy use, it knocked on the door of the Nevada-based company and became one of its biggest clients, marking a major step forward for the fledgling industry.

While many associate hydrogen with the hydrogen bomb and the horrors of nuclear proliferation during the Cold War, the safety of hydrogen fuel is on par with that of solar and is ultimately more mobile and efficient, as hydrogen cells produce no emissions whatsoever. Inspired by author Jules Verne's belief that "water will one day be employed as fuel," Ohmium's business is built around a process that splits water molecules and stores the energy released by the protons in refrigerator-sized energy units that can provide the power of 10,000 EV batteries. It's a development that appealed to venture capitalist and Ohmium investor Ahmad Chatila, who is recognized as a visionary strategist for a global clean energy economy.

With no carbon footprint, green hydrogen can potentially be used to power factories, cars, and eventually homes. "I've been saying for a while that the world should shift to more power from liquid, especially as companies like Tesla have come into the picture," he told us of the industry, currently led by oil and gas, that makes up two-thirds of the worldwide energy market. "The energy density of green hydrogen can become a viable fuel source, especially as the solar and wind costs needed for production have plummeted. Soon we'll be able to power cars. That means that the hydrogen market can be massive. Green hydrogen energy is nothing short of historic."

Ohmium was founded with a mission to create innovative products and services that will enable a sustainable way of life. And by focusing on green hydrogen, the company is fulfilling its mission to create a sustainable world while providing industrial, transportation, and energy projects with a modular, repeatable, and highly efficient model for hydrogen energy cell production. Basically, you take sunlight. You put it in water. You get hydrogen. It is as green as green can be, and yet people don't talk about it outside a very small circle.

"We're going to be the ones who engage in the ecosystem of making hydrogen for all the parts of the economy that don't get solved by electricity and batteries. That's our business model," CEO Arne Ballantine told us of Ohmium's mission to open another path to a post-carbon fuel economy and, given its new client acquisition, make India the hydrogen clean energy hub of the world.

Again, as in so many of our Titanium Economy examples, we see a scalable, practical platform for complex technology production. "One of the things we've done—and there are lots of smart people working on how to make hydrogen—is to do things differently," Ballantine explained. "We've created an architecture, essentially a platform, where we've got a modular way to approach hydrogen generation. Just as you'd see in a data center containing racks and racks of servers, or at a solar farm littered with so many panels, or even in the many tiny batteries inside a Tesla, we're generating hydrogen the same way, but in a way that links up with the capabilities of the Indian economy. Imagine if the thing we were making was the size of a nuclear power plant. Instead, our units are extremely easy to transport and easy to store. So when we talk about creating a hub for green hydrogen generation, we really do have a model to do that."

For Ohmium advisor Pashupathy Gopalan, hydrogen has the ability to emerge as the green fuel of the Titanium Economy. "It's the birth of a new industry, and the timing was quite opportune because, obviously, the world has all started talking about decarbonization," he said.

There Is No Planet B

Aldous Huxley observed, "Facts do not cease to exist because they are ignored." Put more plainly, when it comes to the global climate crisis, there is no Planet B.

By 2050, the anticipated global cost of adapting to climate change in developing countries could reach $500 billion, with the potential to wipe out as much as 18 percent of global GDP. Industrial subsectors with low- and medium-temperature heat requirements, such as construction, food, textiles, and manufacturing, need to accelerate the electrification of their operations quickly. All told, they need to electrify at more than twice their current level by 2050, from 28 percent in 2016 to 76 percent in 2050.

Fortunately, for the first time in history, world leaders and business leaders alike are aligning to reduce the impacts of climate change as public opinion has shifted toward making wholesale changes. This includes the subset of investors, consumers, and state and federal governments putting new pressure on manufacturers to adhere to stricter environmental standards. Investors are putting more money into climate-conscious companies, with almost $90 billion invested in companies developing climate technology in the year leading up to June

2021—an increase of 210 percent from the year prior. Consumers are also much more aware of the environmental impact of their purchases than they have been in the past. Meanwhile, states are passing laws establishing stricter and stricter limits on greenhouse gas emissions, and federal regulators are also taking action. In March 2022, the SEC proposed a rule on climate-related disclosures to help investors better understand the climate impacts a company may be facing, as well as how its board and management teams are navigating those impacts. The writing is on the wall: manufacturers will be rewarded for investing in sustainability and held accountable if they do not.

The good news is that manufacturers are rising to the occasion across the board. In the past decade, American industrial companies have reduced their carbon footprint by 12 percent, and many are working to accelerate their efforts. For over a decade, many companies have requested that their suppliers reduce their greenhouse gas emissions, with some going so far as to ask suppliers to publicly report their emissions. Many companies are embracing innovative technologies to upgrade their equipment, increase efficiency, and replace fossil fuels with alternative renewable energy sources. These moves are both improving the sustainability of companies and enhancing productivity in a positive feedback loop that will drive the acceleration of efforts.

Rewards are also earned through appealing to increasingly environmentally conscious customers, as well as via attracting the next generation of innovators and problem-solvers to jobs in companies that are leading the way. The global climate crisis is a major concern for younger people, with half of this generation of students reporting that they want to work in an environmentally sustainable business. College counselors report that a recent surge in students pursuing environmentally related degrees just keeps gaining momentum. This is yet another competitive advantage for Titanium Economy companies, so many of which are seizing the day with sustainability initiatives.

For many in Gen Z, purpose outweighs profit. Luckily for them—and us—we can do both.

Chapter 10

The Titanium Disruption

In March 2020, a mass email went out from the state of Illinois pleading for help in finding Personal Protective Equipment (PPE). The appearance of Covid-19 had made the masks, gloves, and gowns used by healthcare professionals and other workers one of the most valuable commodities on the planet. Every government entity, from the federal and state level down to the smallest municipality—not to mention hospitals, clinics, public agencies, retailers, and just about every other organization that had any public-facing function—was left scrambling to purchase PPE only to learn that there was no available inventory. Like so many other manufacturers, PPE suppliers had years ago relocated their operations outside of the United States. In fact, just before the Covid-19 pandemic struck in early 2020, China alone was producing half the world's supply of PPE. By March of that year, it had prioritized its own needs in the face of the pandemic.

One person who got that Hail Mary email from the state of Illinois was a self-styled mover and shaker living in Chicago named Jeff Polen. Realizing the urgency of the problem, he immediately combed through his Rolodex and started dialing. That hustle paid off. Polen connected with an associate in China who worked in a textile factory that had recently transitioned to sewing masks and gowns. But the stock it had was mostly spoken for. Nevertheless, Polen's contact took up the challenge and went from factory to factory, piecing together commitments for the use of assembly lines during downtime here and there. In all, he was able to guarantee the production of 1.5 million N95s, the highly protective masks that have been shown to block 95 percent of the small particles, including Covid-19, that few of us had heard of before 2020.

We relay this story not for its satisfying ending—the state of Illinois got its potentially life-saving shipment a month after sending out that desperate email—but as a cautionary tale about the importance of domestic manufacturing. On balance, the United States is the biggest net importer of PPE. We're also the world's biggest net importer of ventilators. That combination made us vulnerable in early 2020 and continues to put us at risk for when the next major health crisis hits.

Our concerns, however, extend beyond these critical products. The Covid-19 pandemic exposed the vulnerability of the supply chains for a wide range of goods, from cars to building materials and a host of products that people count on in their daily lives and that businesses need to operate. The ability to meet our country's enormous needs cannot remain at the mercy of another country's players, priorities, and capacity, nor does it need to. As we've shown in the preceding chapters, manufacturers can make domestic operations thrive.

As we contend with a period of unprecedented disruption, with the Covid-19 pandemic being only one source of upheaval, we can seize the moment to accelerate the growth of the Titanium Economy and realize its true potential. Macroeconomic shifts, technological advances, and changes in capital flows are all driving rapid change, and we see significant headwinds that might impede the growth of the US industrial-tech sector. If we don't move quickly, we will continue to fall behind. But if we learn the lessons of our past and look to our peers in the global economy for inspiration, we can navigate our way forward through those headwinds. The good news is that many Titanium Economy companies are aware of the myriad disruptions they are facing and are acting fast.

The first source of disruption, macroeconomic shifts, is reshaping the business environment. This includes geopolitical tensions, which are making the future of the longstanding international order unpredictable. Climate change and the shift to net-zero emissions by 2050 would transform the global economy, requiring $9.2 trillion in annual average spending on physical assets for energy and land-use systems, $3.5 trillion more than today. The Covid-19 pandemic not only caused untold misery and lives lost, it also impacted the global economy as we experienced supply-chain strains.

Against the backdrop of these macro shifts, the pace of technological change continues to gather steam. Quantum computing, while still in its infancy, promises to be a game-changer in fields from cryptography to aviation, data analytics, and many more. This new breed of computers can analyze much larger volumes of data and tackle much more complex problems than even the most powerful supercomputers today. To illustrate the potential, if you wanted to find a single item on a list of one trillion, and each item took one microsecond to scan, a classical computer would take about a week to accomplish the same task that a quantum computer would complete in one second. This new superpower is set to deliver more than $1 trillion in potential value by the mid-2030s.

Meanwhile, cloud-based computing, paired with artificial intelligence and advanced manufacturing applications, has provided a galaxy of applications and continues to grow rapidly. With 125 billion connected IoT devices expected to be sold by 2030, and with the expansion of 5G and the coming development of 6G telecommunication networks, industrial robots—already numbering more than 2.7 million—will increasingly support advanced material development, 3-D printing, and other applications that are so crucial to the Titanium Economy. This connected high-speed 5G infrastructure will drive further technological advances in electrified transportation, energy storage, and smart grids, as well as the development of smart cities that are much more energy efficient and sustainable. But, of course, the pace of these developments will depend on the volume of investments made in them.

The third source of disruption is changes to the capital supply, or the amount of money available to invest in assets around the world. This capital supply invested in various asset classes is determined by the risk-return curve. Typically, investors expect higher-risk securities like equities to deliver higher returns while lower-risk securities like bonds typically deliver lower returns. Think of it like this: The odds of Tom Brady leading his team to win the Super Bowl in the first two decades of the twenty-first century at the start of each season would have been higher versus a rookie quarterback doing the same. Therefore, bettors would expect a lower return after putting money on Brady given those higher chances. In investment circles, this same concept—albeit of an investor's expectations on how much return to accept given the underlying risk of the asset class—is called the risk-return curve. Over the past three decades, the global capital supply has increased six-fold, from $118 trillion in 1990 to more than $700 trillion in 2018. Moreover, every category of investment has experienced this growth, both in public and private equity and debt. However, this growth has not been even across all asset classes. In recent years, corporate valuations have increased to all-time highs, and interest rates have been lowered by central banks across the globe to spur economic growth. These trends have fundamentally altered how investors value risk. They've been expecting higher returns than ever before from higher-risk assets, which means that the typical risk-return curve has steepened. That, in turn, has fueled more flow of money toward high-risk assets in a self-perpetuating cycle.

We expect this trend to continue, in part due to a funding gap in pension funds—one of the largest sources of investment capital. Twenty years ago, US pension funds were fully backed with respect to obligations to their constituents. In other words, they had enough money on the books to pay all the pensions they were committed to. But they now have an aggregate funding gap of nearly 30 percent, meaning they only have enough to fulfill 70 percent of their pension obligations. This is pushing the funds to put more of their assets into higher-risk investments, hoping to earn higher returns with which to bridge the gap.

There is clear opportunity in the industrial sector. With returns premised on actual performance—rather than primarily on investor optimism as has been the case in other sectors such as technology and telecommunications—the industrial sector has delivered more reliable shareholder value over the past decade. If institutional investors and the investment community broadly seize the opportunity, the Titanium Economy in the United States can leapfrog past the competition abroad.

Making that shift in investment is critical because the United States is in a race to build preeminent manufacturing competence, and unless we act, we face real danger that they will leave us behind. Below, we discuss specific ways to mitigate the effects of the headwinds threatening to impede our progress and take advantage of tailwinds that can boost it forward. We need all hands on deck; our prescription for change involves the entire ecosystem of the Titanium Economy, including leaders within companies, the investment community, and policy makers.

Prescription 1: Industrials can leverage the Titanium Playbook to lead their micro-verticals.

We believe in the potential to make the entire industrial sector Titanium strong, with hundreds if not thousands of Teslas, HEICOs, and Middlebys. But achieving this will require corporate leaders beyond those guiding the current twenty to thirty superstar performers to transform their enterprises into high-technology-driven companies.

In our experience, the best Titanium Economy companies drive value by leveraging technology to use data more intelligently and to create products that add substantial value for their customers, assuring they will pay more for them. If you're manufacturing a mixer or a pump, for example, you can add smart sensors to track the health of the equipment and diagnose maintenance issues more proactively. You can then share that data with your customers, which helps them run their businesses better, making the value you've generated for them clear. There's nothing magical about margins; you raise them by improving your products and performance to provide the extra value customers will pay a higher price for. We're not talking only about revenue growth. If companies now have a 5–10 percent EBITDA, we'd want to see them reach 15–20 percent EBITDA. Today's wave of new technologies offers unparalleled opportunities to achieve such revenue increases by transforming product infrastructures as well as products themselves. Applied properly, data analytics can help make better decisions not only on product development but on pricing, sales, costs, and plant operations.

Titanium Economy companies that are already achieving impressive earnings have no reason not to aim even higher, as Emile Chammas of Sealed Air told us they've done. "We launched an initiative called Reinvent SEE to transform the company when we already had the high margins and one of the highest multiples in packaging," he said. "We don't benchmark to others, but to ourselves—where we are versus world-class. When we announced the project in December 2018, some concerns were voiced about pushing too much. Nevertheless, we announced a $200 million margin improvement plan by the end of 2021, and the market reacted favorably. Over the next eighteen months, we systematically went after it and exceeded the targets we set before the deadline. We streamlined processes and improved the interactions we had internally, as well as with our customers and suppliers. We also drove organic growth while using mergers and acquisitions to move from a product company to a world-class, digitally empowered company, acting like a startup to disrupt the markets we serve, our industry, and ourselves."

Like at Sealed Air, the transformation can happen in a short time frame. In our experience, companies that give themselves five to ten years to adapt end up never achieving their goals. Conversely, if your mindset is to change performance in an accelerated fashion, giving yourself eighteen to twenty-four months to get to the next level of performance—and we mean a transformational rather than an incremental jump—you'll find yourself much more vigorously adopting new technology. And the process shouldn't stop there. Once a company has fixed its core operations, it can aim to become a segment of one and lead its micro-vertical.

Focusing on the core business is key to success here. If you broaden beyond that, you run the risk of losing the intimate knowledge of your customers and the best ways to offer them more value. If you're HEICO, you're an aerospace parts company. If you're Graco, you're a fluid-handling company. This is because what you do, and for whom, is so clear that you don't need to spend a lot of money on sales and marketing. As for R&D spending, a majority of it is rigorously dedicated to innovations you know will matter—not to wild goose chases. Leveraging M&A opportunities to gain scale can also be an effective strategy for becoming the number one or two player in a microvertical, as HEICO and Middleby have achieved.

The playbook is proven and is available to any industrial company.

Prescription 2: Industrial leaders can do a better job of getting their companies valued at what they're worth.

In the last decade, most sectors gained in value simply because of investor expectations improving. Across sectors, no matter what company we looked at, there has been an expansion of the multiple of a company's valuation over its revenue, or EBITDA (the average benchmark for a sector) has increased because investors have been willing to pay more. In fact, on average, multiples have increased nearly 80 percent since 2010, whereas the actual profits of companies decreased on average by 30 percent. Take telecommunications, for example. Two-thirds of the increase in valuations has come from investors valuing the companies more highly than they did ten years ago. If you're a shareholder in a company today, in other words, nearly 90 percent of the appreciation of your shares' value over the last decade came simply from multiples being higher.

What's more, there is a very limited correlation of multiples with actual returns. This means investors have been expecting better performance, but that has not been substantiated. In fact, the correlation is less than 10 percent, which is incredibly poor. The growth rate of companies also does not correlate well with multiples. Based on the data we've analyzed, we can say that less than 40 percent of the expansion in multiples can be explained by actual performance improvement.

When we said we saw this phenomenon across all sectors, we should have said all but one. In the industrials sector, only one-third of shareholder value in the past decade has come from higher multiples while two-thirds has been due to actual gains in performance in either revenue or margin expansion. The irony is that multiples in the sector are the lowest across the economy because investors simply aren't as excited about industrials. But if investors looked under the hood of companies in the industrial sector, they'd find at least forty companies with tremendously attractive characteristics.

So how do investors identify the next HEICO, Graco, or Middleby? One significant factor we've identified is the quality of revenue. To judge that, we ask a number of questions: What markets does the current company play in? What type of customers are they serving? Are they serving the winning customers, or are their customers not doing so well? Do they have an intimate relationship with the customer, allowing them to develop products together, or are they just another supplier? What kind of intellectual property do they bring to the market? How do they monetize their products? Do they have ongoing relationships with customers, or is each sale a one-off? Another factor that's important to valuation is corporate oversight. What is the ownership structure? What kind of CEO, and what sort of board, supports the company?

To impress the appeal of investing in these firms upon analysis, industrial-tech companies need to play a better game of making their case, becoming fluent in the language of the investment community.

Prescription 3: Investors can change the way they identify and price industrial diamonds in the rough.

In turn, investors can develop their understanding of the industrial-tech sector. First, they can endeavor to find investment opportunities not by looking at the sector as a whole, but by breaking it down into micro-verticals and then into the players within those micro-verticals. Then they can assess each company as a member of that unique niche. That way, they will be better able to understand their valuations: Are they reasonable? Are they overpriced? Are they underpriced? In short order, investors will be able to identify companies with reasonable valuations and multiples that are not very high, which could therefore present significant upside potential relative to their current profitability and operating level.

In evaluating that potential, investors also can determine the quality of revenue for companies. A low multiple isn't sufficient for assessing upside. Within the set of companies with low multiples, a subset has higher-quality revenue in terms of growth, operating leverage, and profitability. These companies with stronger fundamentals that form this subset are the diamonds in the rough to focus on. And within that group, the companies with the greatest potential have the best prospects for applying new technology to boost performance.

In addition to this discovery work, investors can become more engaged with these companies. They can't be financial sponsors only. They have to become involved in helping companies drive change. A great model for this is Flagship Pioneering, a venture capital investment firm focused on generating breakthroughs in human health and sustainability. To deliver breakthrough innovation, Flagship has developed an origination model in which it is both the founder and funder of portfolio companies. These companies emerge from a structured process to generate innovation at scale that begins in its wholly owned Flagship Labs. Each portfolio company develops a platform technology with the potential for breakthrough innovation that is far beyond incremental ideas traditionally pursued by the large incumbents. To help each portfolio company achieve its full potential, Flagship maintains an ecosystem with significant scientific, technical, and financial resources. Through such an engaged approach to company origination, Flagship Pioneering has leveraged its ecosystem to support its portfolio companies and bring us world-class, innovative companies like Moderna.

The bottom line is that investors can become more granular about how they look at the opportunities among industrials, and when they do take the plunge, they can go beyond capital development. If they really engage, if they spend time with their industrial investments and help them grow on the sales side—for instance, by making introductions to their own customers or helping to run the business better and more efficiently through digital resources—then the rewards will speak for themselves.

Prescription 4: We need to think big on talent before the momentum of industrial innovation stalls.

The skilled trades, such as electricians, plumbers, carpenters, and HVAC technicians, have been in a state of crisis for close to a decade, and this has been exacerbated by the Covid-19 pandemic. Only seventy out of one hundred positions are being filled, translating into 2.4 million unfilled positions in manufacturing and an estimated $2.5 trillion of lost economic value over the past ten years due to the labor shortage. And the problem is expected to get worse because more people are currently leaving the trades than joining them. Meanwhile, as more companies look to move their supply chains and production back to the United States, demand for trade workers will grow. We need to consider reforms to aspects of our education system.

First, career and technical education (CTE) can be reinvigorated in high schools, with a well-structured entrance process for interested students. CTE was long an integral part of the US education system, but between 1990 and 2009, CTE credits earned by high schoolers decreased 14 percent. Over half of CTE-related jobs available in the United States are in four fields: business management, manufacturing, hospitality, and marketing. However, these fields account for only a quarter of CTE course concentrations today. While the past decade has seen a resurgence of interest in CTE as states instituted a patchwork of new laws or policies regarding the programs, we can support this on a national level and accelerate the resurgence of CTE for the next generation.

Second, there is an opportunity to create a national Apprenticeship Institute that works to elevate the importance of apprenticeships in American culture to the level of university degree programs. The federal government could help create such education and apprenticeship programs through the introduction of supportive policies and the provision of funding. Germany's dual training apprentice system is a leading model to consider, with apprenticeship training provided by companies and academic classes alike. Industrial firms in the country see the training they offer as the best way to acquire skilled staff. As for students, more than half of high school graduates go into the system, with the apprenticeships being just as celebrated as college degree programs.

The Apprenticeship Institute would be tasked with launching a national campaign to rebrand apprenticeships and vocational training to combat the negative perceptions in the American psyche. Singapore is a shining example of a successful national branding campaign—today, 90 percent of their bottom-quartile achievers graduate from a technical institute and secure well-paying jobs, whereas the US bottom quartile is at the highest risk to drop out of high school.

A third necessary component to consider in any reform effort is a national strategy for providing the upskilling that workers will need as the technological disruption of the Fourth Industrial Revolution proceeds. As machines increasingly take over manual tasks, companies and the government can assist trade workers in adapting to the new roles they'll be asked to fill. This reskilling of the workforce to work in increasingly automated plants won't be easy. Workers will be required to learn new skills such as writing software, even at the lowest levels of production, and contribute to designing better products and conducting data analytics. They'll need to become better problem-solvers, making use of the tremendous amount of data that will be at their disposal. The ability to think critically about problems and communicate with coworkers will be required of every factory employee of the future.

SkillsFuture, and its sister agency WSG in Singapore, is a good model of a skill-building program. It's open to all citizens, regardless of background. Mid-career individuals are supported and acquire relevant skills through training directly with reputable companies, which is something the United States can replicate. The federal government could also rewrite unemployment provisions to support part-time work and technology transitions through shared government-employer responsibilities, as Germany does, which would enable workers to stay employed and productive while they learn new skills.

Prescription 5: We can make the Fourth Industrial Revolution a national priority and build the infrastructure to support the adoption of industrial technology.

As we have pointed out throughout the book, the technology disruption is well underway. Quantum computing, cloud computing, artificial intelligence, machine learning, advanced materials, and 3-D printing are already making the industrial sector more competitive, more efficient, and better at serving the needs of its customers. The growth of the internet of things, robotics, and automation is expected to be 20 percent every year for the next five years. Meanwhile, we are witnessing in real time the evolution of the use of electric vehicles, drones, and autonomous driving in the industrial sector. The United States is more than capable of owning these sectors. Failure to do so could create significant challenges in the future.

Case in point: In the lithium-ion battery industry, the United States only accounts for 10 percent of global manufacturing capacity, with nearly 60 percent claimed by China. As lithium-ion batteries are key to electric and autonomous vehicles, if we continue to lag in the space and become dependent on foreign technology and capacity, we will endanger our ability to chart our own manufacturing destiny.

China has gotten so far ahead of the curve because the government is dedicated to taking the lead, providing the funds to make sure its industrial base is filled with top-notch technology companies and not just efficient factories. As the window of competitiveness closes, US industrial executives now face the reality that they can no longer afford to be non-tech companies.

As evidence of how quickly competition is evolving, just look at what happened in the semiconductor manufacturing field. For decades, the United States was the leader in the sector. Over the past three decades, however, Taiwan and South Korea have gained dominant market shares as the US share has declined. As we've recently experienced, the supply shortage of semiconductor chips impedes not only our ability to deliver computers but also everything from cars to basic consumer goods.

Industrial executives must realize that we have maybe five years to close these gaps. That's it. Whether it's in lithium-ion batteries, semiconductor manufacturing, electric vehicles, or AI, sitting back simply isn't an option because if we don't do it, those across the Atlantic or Pacific will. And when we say that the cost of waiting will be tremendous, that's not simply talking about opportunity cost. Sure, jobs will be lost. But there will be a cost to be paid in hard currency by everyone in society too.

Here again, the federal government could play a role. One idea that can be considered is the appointment of a chief industrial technology officer—a cabinet-level position—in charge of steering our national innovation strategy and policy to support the industrial sector. This is an idea that has already been proposed. In a bipartisan effort, Congressmen Tim Ryan (D-OH) and Tom Reed (R-NY), sponsored HR 2900, from the 116th Congress, which called for the creation of a US Chief Manufacturing Officer (CMO) in the Executive Office of the President to coordinate "manufacturing-related policies and activities across government" and "develop a National Manufacturing Strategy to establish a clear path for growth in the manufacturing sector." With the right national industrial-technology strategy, we can chart a path for securing preeminence in artificial intelligence, robotics, and other new technologies, focusing funding and investment in those areas. We would be keen to see the public and private sectors come together to create an organization, the American Industrial-Tech Institute, charged with pursuing these initiatives, which would help prioritize and elevate the groundwork laid by Manufacturing USA. This would constitute a centralized and expanded set of public and private institutes focused on developing academic innovation and then transferring those advances to the relevant industries. This bridge between academia and the private sector would quicken the pace of getting innovations out to industrial players—and then to the market—in a scalable fashion.

Lastly, fiscal policies and investment in strategically determined areas could support stakeholders and unlock real potential. We can look to South Korea in this regard, where tax relief is provided to companies making capital expenditures for investments associated with research and development. Drawing from lessons in other countries, there are a number of big ideas to consider. For example, the US government could consider "American-made" government bonds that are lower on the risk-return curve. Another approach we've seen in other countries is the creation of a national development bank that issues bonds on private capital markets, distributing them to small- and medium-sized enterprises.

There is no shortage of ideas and opportunities available to support the future of a strong industrial-tech sector.

The Time Is Now

In this period of massive disruption, we find ourselves at an inflection point, and we as business leaders, investors, policy makers, and citizens have a clear choice between two paths. One is business as usual, which could see the United States falling to the third-largest economy behind China and India. In this scenario, the United States battles rising inequality while low-wage jobs outside of industries see the largest growth. The other path leads to a future in which the United States regains its preeminence as a world manufacturing superpower, with Titanium companies springing forth in communities all across America, igniting Great Amplification Cycles as players in vibrant Titanium hubs and strengthening the socioeconomic backbone of the country.

Action is urgently needed, and the payoffs for first-movers will be extraordinary. Companies that act now, and decisively, will have the greatest chance of becoming one of the players within their micro-verticals. On the flip side, the cost of inertia will be high. When more and more investors start taking a closer look at industrial companies, the likelihood of discovering an undervalued Titanium Economy company will decrease. The argument for urgency holds for public policy as well. Every major manufacturing-dependent country is currently putting together proactive industrial policies to educate its workforce and build the infrastructure needed to achieve global leadership in the industrial-tech sector. In addition, the public policy changes that are needed in education will have a long lead time for building up the necessary workforce. The United States is more than capable of leading the pack.

The United States must act, as the cost of delaying will be severe. Because we believe that a blueprint for action is needed to galvanize the required changes, we're closing the book with thirty ideas in the spheres of innovation, education, and government. We must work together, whether you're a company executive, a policy maker, a factory worker, or a student considering career options. Our immersion in the companies and communities of the Titanium Economy assures us of a better future for all. We simply need you to help bring the vision to life. The time to act is now. We hope we've inspired you to join the cause.

Thirty Ideas for Innovation, Education, and Government

Innovation

- **Consider the appointment of a chief industrial technology officer** and a dedicated body that defines strategic national manufacturing priorities, consults on industrial-technology matters, drives public-private partnerships, and supports subsidies and incentives.
- **Explore establishing a national technology strategy defining** AI, robotics, and Industry 4.0 technologies as priorities, and focus funding and investments in those areas.
- **Prioritize** and elevate the groundwork laid by Manufacturing USA by considering the establishment of an American Industrial-Tech Institute, a centralized set of public-private institutes focused on transferring academic innovation to businesses for the purpose of new product developments, technology enhancements, and efficiency gains.

- **Designate a national network of geographic "Innovation Hubs,"** areas that co-locate industry, academia, and finance to improve collaboration and innovation output, with an opportunity to rebalance attention to areas beyond Silicon Valley and other coastal hubs.
- **Prioritize industrial-tech investments to shift attention and value to the sector** and shift the investment glut away from standard tech, recognizing the dynamism and multiplier effect of US manufacturing.
- **Incentivize increased employee ownership and associated business models** by encouraging companies to implement programs that align employee interests with shareholders, increase employee engagement, and improve business culture and innovation.
- **Build tomorrow's infrastructure today** through large-scale projects that support technology and clean energy solutions for industry and the public.

Education

- **Reinvigorate career and technical education (CTE) high school programs** and develop a robust and well-structured entrance process for students who are interested in skills-based education.
- **Create opportunities for free post-secondary college, vocational, and master's education programs** to develop the most advanced, skilled workforce in the world and increase opportunities for marginalized groups.
- **Increase enrollment in skills-based post-secondary education** to train students with technical skills that are in demand by employers.

- **Create "Master" school hubs,** following the example of Germany's Meisterschules, a national network of partnered programs between education and industry with focuses on specific areas of expertise, including AI, additive manufacturing, robotics, and more.
- **Elevate internships and eliminate unpaid internships** to position them as an onboarding mechanism for apprenticeships that incentivizes skill building in early tenured employees.
- **Create a national "Apprenticeship Institute" network** across all states to elevate the importance of apprenticeships in American culture to the level of university degree programs.
- **Launch a national campaign to rebrand apprenticeships and vocational training** to combat the perception that they are inferior to university degrees.
- **Develop a national strategy around upskilling to "focus on the future"** by creating leading-edge skill-building and upskilling programs that will return long-run value to employees and employers.
- **Scale and incentivize efforts to connect job seekers and employers** from the beginning of education to employment, engaging everyone in mentorship from the start of their careers.
- **Incentivize "pay for success" training programs,** potentially with traditional community college and university systems, in which program fees are partly based on whether a student is hired.
- **Develop workforce training programs, specifically focused on upskilling low-wage earners and underrepresented groups** to close the gap in skills needed by future industries.
- **Create a nationally recognized credential verification resource** that can help job seekers prove their skills to employers by providing a centralized verification system.

Government

- **Assess the opportunity for a new National Development Bank or Corporation** to issue bonds on private capital markets and distribute them to small- and medium-sized industrial companies in the United States.
- **Consider developing "American-made" government bonds** that may be lower on the risk-return curve.
- **Continuously improve the national unemployment scheme,** including the potential to support part-time work (and technology transitions) through shared government-employer responsibility and with programs to allow workers to remain employed, productive, and learning through economic slumps.
- **Evaluate investment opportunities in the American "semi-periphery"** (deindustrialized areas of the continental United States), leveraging public-private partnerships, to create basic inputs for an emerging industrial-tech sector.
- **Support efforts to bring supply chains closer to home, including potential incremental funding** to strengthen their resiliency and create jobs.
- **Repair public works and infrastructure** to increase demand for an underemployed labor pool and to smooth the transition toward a more highly skilled industrial workforce.
- **Follow best practices in aligning capital expenditure** with areas identified as national innovation priorities.
- **Pursue options to help stimulate capital investment** and improve long-term competitiveness, such as tax incentives to support Industry 4.0 technology adoption.

- **Help drive R&D investment in innovation** by aiming to match the levels seen in Germany and South Korea, for example.
- **Increase US attractiveness for investment** through a variety of measures, which could include competitive corporate tax rates.
- **Promote flexibility and adaptability in trade** while reducing barriers to trade.

Charts

The industrial sector can be divided into 90+ micro-verticals

Titanium Economy companies can be found in each micro-vertical

Segments	Number of US-traded public companies	Number of micro-verticals	Examples of segment micro-verticals	
Automotive and ancillaries	105	12	Agricultural and farm machinery; Auto parts and service	Motor vehicles and passenger cars;
Aviation, aerospace, and defense	46	3	Aircraft/aerospace parts and equipment; Aviation and defense OEMs	Ordnance and accessories
Electrical components and equipment	32	5	Electric lighting and wiring equipment; Electrical storage and distribution equipment	Solar modules; Motors and generators
Electronic components and equipment	156	21	Automation and control; Lasers and related devices	Communications components; Test and measurement sensors
Home and building products and technology	101	12	Fire and security equipment; Building-material distributors	Cooling and heating equipment; Building exterior products
Industrial diversified	23	2	Industrial diversified; Industrial-focused PE	
Industrial machinery	81	11	Industrial tools; Metalworking machinery	Fluid handling (pumps, valves, meters); Food-processing equipment
Industrial materials	17	3	Industrial gases; Specialty/engineered materials	
Industrial services	94	15	Industrial-waste solutions; Industrial engineering, procurement, construction	Waste collection and disposal; Engineering services
Industrial trading and distribution	33	6	Wholesale trade-durable goods; Container/marine-equipment rental and leasing	Aircraft-equipment rental and leasing Services-equipment rental and leasing
Total	**688**	**90**		

Source: McKinsey report (November 2020), Value creation in industrials

Micro-verticals are clusters of similar companies defined by product & end market

Titanium Economy companies seek to lead in their micro-vertical

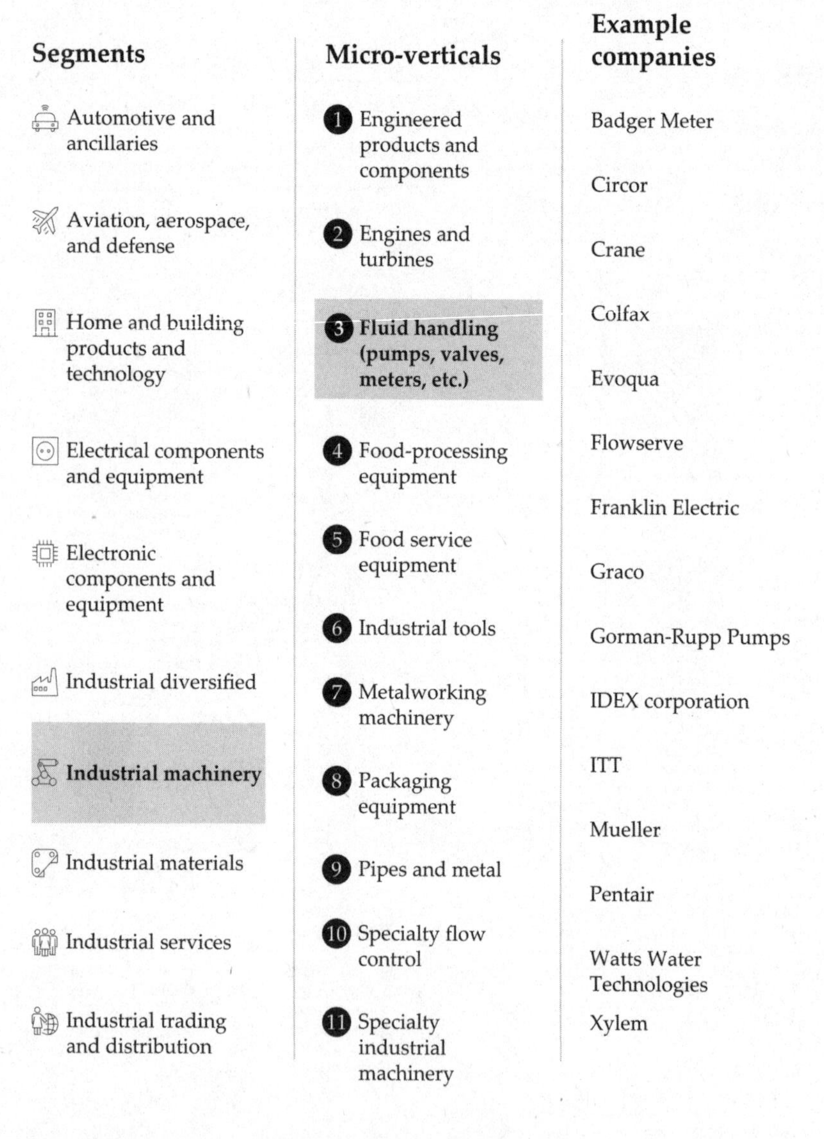

Source: McKinsey report (November 2020), Value creation in industrials

Most industrial companies are private, and more than 80 percent of listed companies are small or midcap

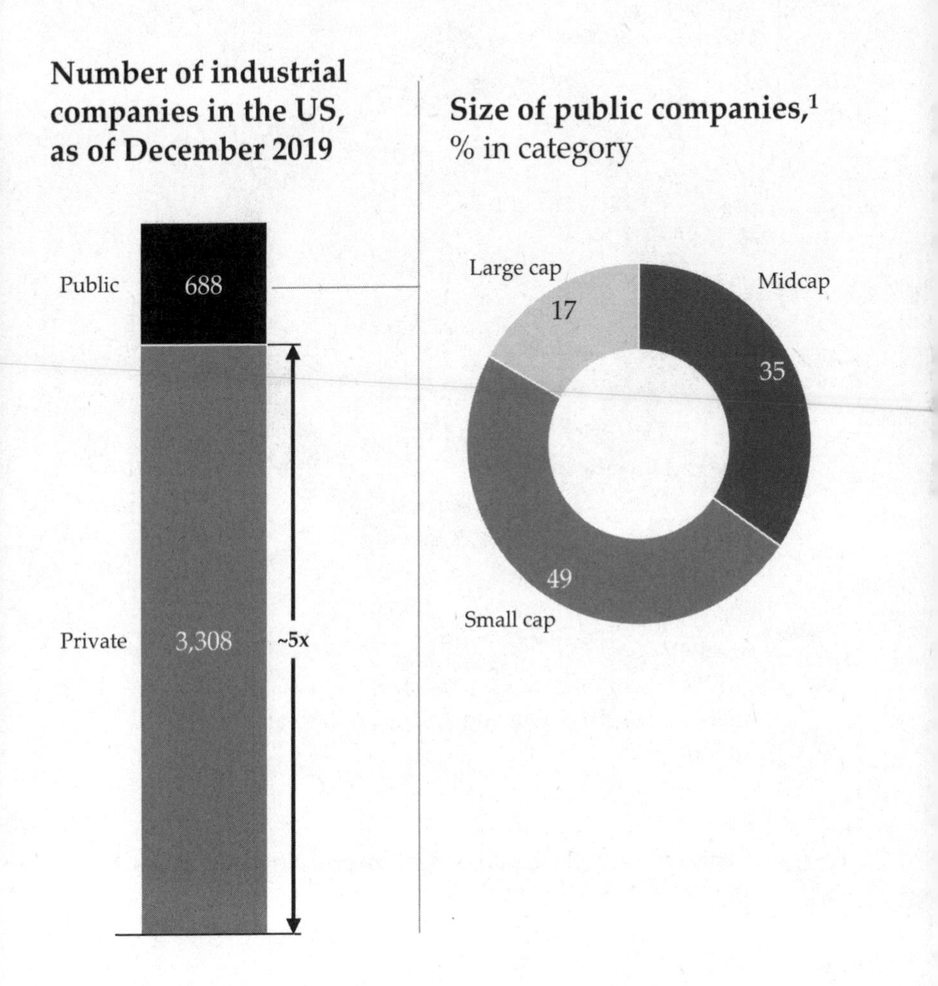

1. Capitalization defined based on 2019 revenue; large cap, >$5 billion; midcap, $1 billion–$5 billion; small, <$1 billion. Figure does not sum to 100, due to rounding.

Source: McKinsey report (November 2020), Value creation in industrials

The Titanium Playbook

Top-performing companies improved their performance by leveraging a three-step approach

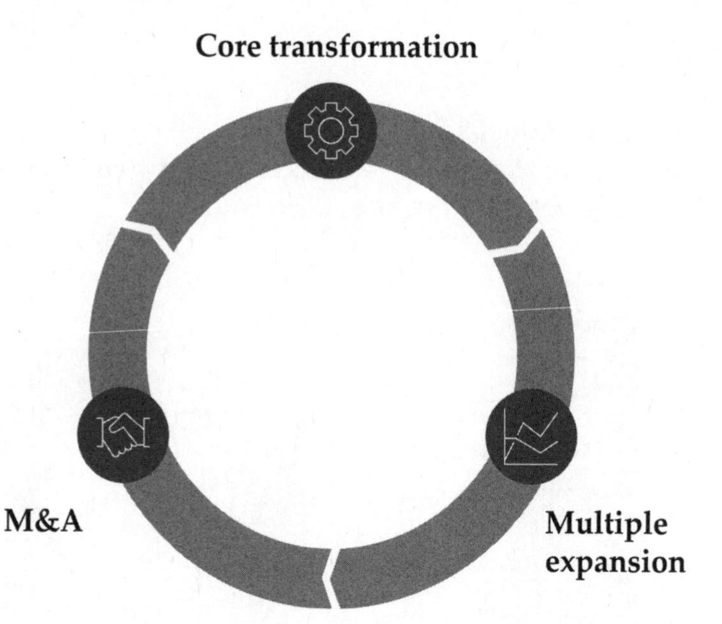

Step 1
Drive core transformation, leveraging tech and data, to achieve margin expansion and growth in an accelerated fashion

Step 2
Focus on multiple expansion to improve investor attractiveness while improving performance

Step 3
Leverage M&A to achieve "segment of one" status and build a platform for future expansion

The Titanium Economy promises a better life for all Americans

The transition of the US economy from manufacturing has coincided with the exacerbation of income inequality

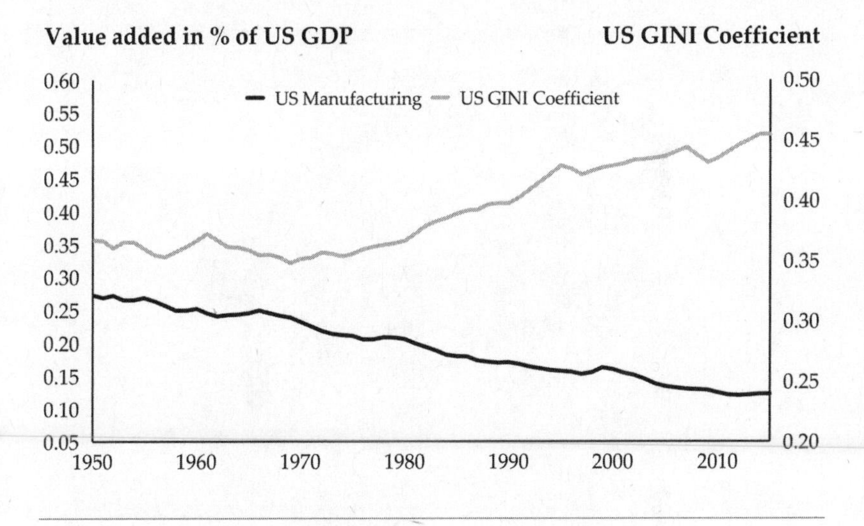

... and it is the highest among G7 countries

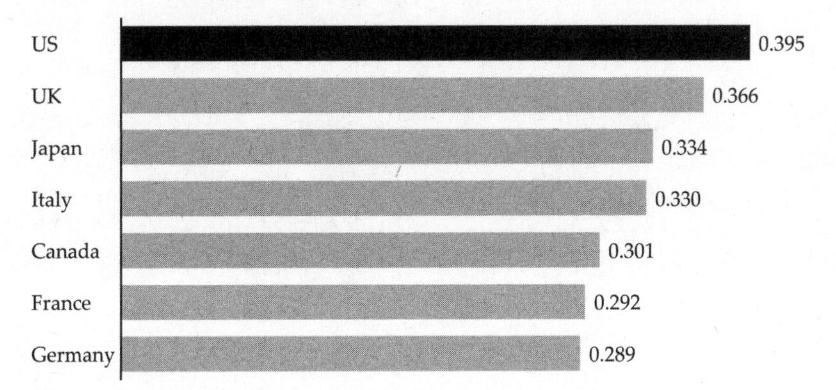

Note: The Gini coefficient is based on the comparison of cumulative proportions of the population against cumulative proportions of income they receive

Source: Pew Research, OECD

Supporting the Titanium Economy can help bridge divides in employment

60 percent of net job creation through 2030 may be in twenty-five urban areas

Projected net job growth in midpoint automation scenario, 2017–2030

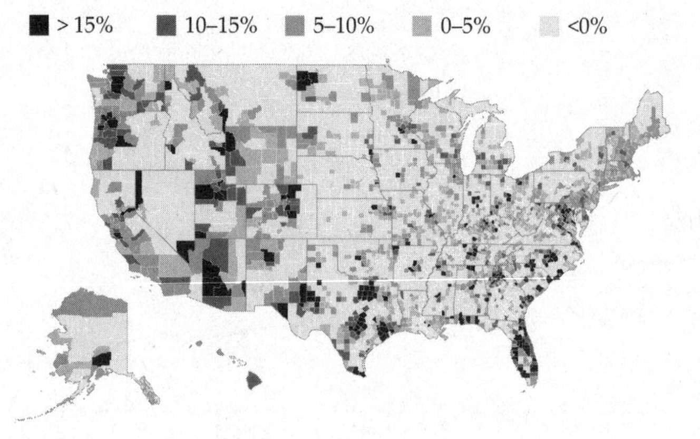

Manufacturing accounts for over 20 percent of employment in 460 US counties—most of which will not benefit from expected job creation

Source: US Census American Community Survey; McKinsey Global Institute analysis

Industrial companies have driven value from true performance improvement instead of expectations

Multiples are the primary engine of shareholder value creation, outstripping economic profit, except in industrials

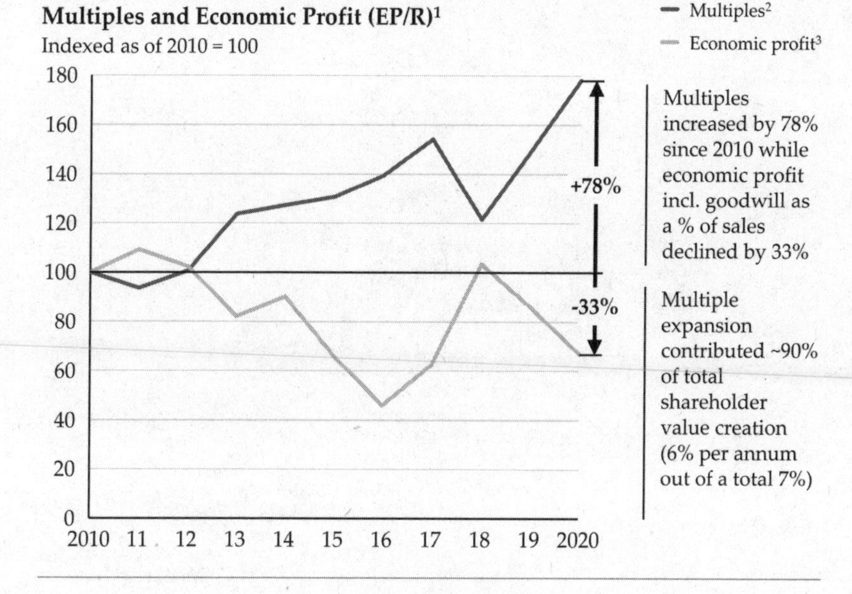

	EP increase 2020 vs. 2010	Multiple increase 2020 vs. 2010	Share of TRS growth due to multiple expansion
Tech and telecom	96%	115%	55%
Consumer	6%	76%	93%
Industrials[4]	72%	47%	40%

Multiples are the primary mode of value creation, except for industrials

1. Based on sample size of ~2,300 global companies
2. EV/NOPLAT calculated based on ratio of aggregate EV to aggregate NOPLAT
3. Calculated as ratio of aggregate economic profit incl. goodwill to aggregate revenues
4. Industrials excluding airlines

Source: Corporate Performance Analytics

Major disruptions will drive opportunities for industrials—businesses, investors, and policy makers can seize these opportunities

Key disruptions: Macro shifts, technology advances, and changing capital flows

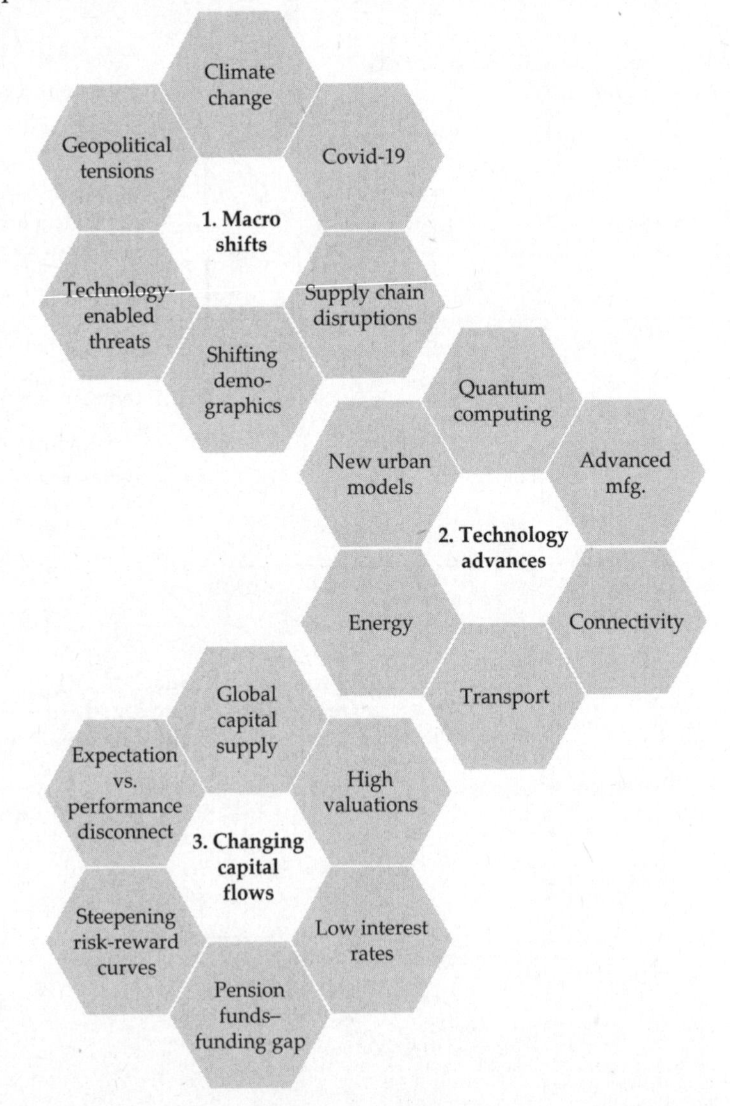

The Titanium Economy can be best supported by key stakeholders working together

A non-exhaustive list of ideas for stakeholders:

Leaders of industrial companies

- Upskill to lead in the disrupted future
- Set aspirations for what success looks like
- Measure and adjust continuously

Industrial companies

- Accelerate transformation to create value
- Focus on multiple expansion
- Leverage M&A to create a "segment of one"

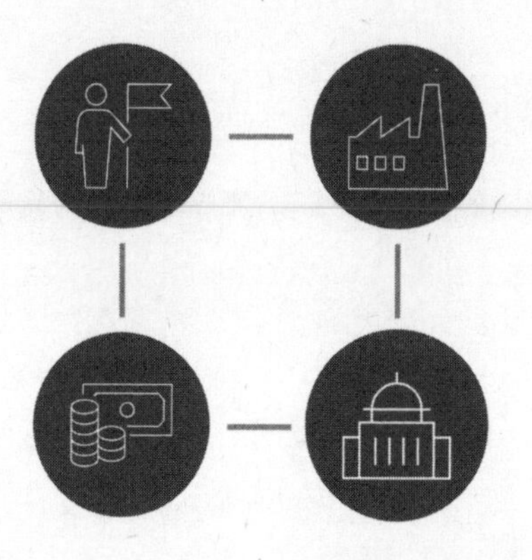

Investors

- Leverage the "X-factor" in investing beyond capital deployment through, e.g.,
  - Building innovation ecosystems
  - Building accelerator ecosystems
  - Hands-on investing
  - Intellectual capital

Policy makers

- Continuously improve education: e.g., Career and Technical Education (CTE)
- Assess innovation opportunity: e.g., a centralized set of public-private institutes focused on transferring academic innovations to market
- Follow best practices in funding/fiscal policy: e.g., capex incentives, on-shore ecosystem development, competitive taxes

Acknowledgments

We are deeply grateful to the many people who contributed to *The Titanium Economy* and the insights contained in these pages. First, we give thanks to the twenty-five industrial leaders whose stories and experiences find prominent roles in the ideas of the book and to the more than fifty additional leaders who offered their personal stories and perspectives as we uncovered the promise of this vibrant sector. They are role models for so many and stand as evidence of the types of impact that can be realized with the right aspirations.

Next, we want to thank those who helped us conduct the necessary research and analytics, comb through countless pages of interview transcripts, and sort through the array of insights that were needed to truly understand the potential of the sector. The leaders of our research team, Jared Retka and Jacklyn Nagle, who were joined by Rob Konkel, Adam Tappella, Nico Joy, and Riad Hamade, cannot be thanked enough. Without their efforts, this book would not have come to life so quickly and so fully. Our executive assistants, Sandra Sprague,

Barbara Albano, and Danielle Isbell, were a tremendous help and a constant partnership through this process.

We were keenly aware of the need to tell the stories of the industrial sector in a way that would bust the myths that have become so deeply embedded in the American psyche. We wanted to do the sector justice while ensuring a story that was engaging and compelling. Our collaboration with Ken Wells, Herb Schaffner, and Jeff Slate made the book possible and ensured we could achieve the best balance between insight and storytelling. We were expertly guided by our agent, Lynn Johnston. Thank you for all your support on this journey. Thank you to Raju Narisetti, the leader of McKinsey's Global Publishing group, for encouraging us to work with Lynn, and also to Ben Adams, our editor at PublicAffairs, for your thoughtful partnership as we worked to shape our thinking. Assistance from Emily Loose, Jamie Leifer, Clive Priddle, and the dedicated team at PublicAffairs helped to make this publication what it is today. Thank you as well to Tim Ward, Adam Braveman, David Harris, Hana Katen, Cam Henry, and Salvador Gómez-Cólon for your support in this effort.

As a partnership, we are lucky at McKinsey to have many colleagues who are willing to help us bring the best of our ideas forward. First, we would like to thank Bob Sternfels, whose passion inspired us to look deeper at the industrial sector and how it can be a source of holistic impact to create a world of sustainable and inclusive growth. We would like to thank Sven Smit, one of the authors of *Strategy Beyond the Hockey Stick: People, Probabilities, and Big Moves to Beat the Odds*, for his help and counseling during the early days of this effort. We are particularly thankful for our colleagues in the Advanced Industries practice: Kim Borden, David Ebenstein, Inga Maurer, Alex Panas, Shekhar Varanasi, Akshay Sethi, Nidhi Arora, and Siddarth Madhav, who are also leading counselors in the industrial space and provided us with inspiration and ideas that have been central to this work. They, in addition to Osato Dixon and Myron Shurgan, have been strong support throughout this process. Thank you to our many partner colleagues across Advanced Industries and our firm, who are too many to mention by name, for your constant support and partnership.

We are deeply grateful to our families, most importantly our significant others, Rita Anand-Padhi, Sujatha Santhanam, and Priyanka Kher, for their counsel and encouragement through the years of work that were reflected in the writing of this book. Words cannot express how crucial each of your support has been for us over the years. We all have full-time client service roles that kept pace during the writing of this book, and without our families' support, this effort would not have been possible.

Finally, to all of you, we are thankful to you for believing in the promise of the Titanium Economy. We want to ensure that we are doing all we can to guarantee this future is realized, and in that effort, we welcome any ideas and feedback you are willing to share.

References

Chapter 1: Underappreciated, Undervalued, and Misunderstood

EOS. "Liebherr: First Metal 3D-Printed Primary Flight Control Hydraulic Component." 2021. https://www.eos.info/en/presscenter/expert-articles/aerospace_liebherr_ventilblock.

Fraunhofer-Gesellschaft. "Fraunhofer-Gesellschaft Profile / Structure." Updated January 2022. https://www.fraunhofer.de/en/about-fraunhofer/profile-structure.html.

Gartner. "Gartner Forecasts Worldwide Public Cloud End-User Spending to Grow 23% in 2021." April 21, 2021. https://www.gartner.com/en/newsroom/press-releases/2021-04-21-gartner-forecasts-worldwide-public-cloud-end-user-spending-to-grow-23-percent-in-2021.

Grand View Research. *3D Printing Market Size, Share & Trends Analysis Report . . . 2021–2028*. May 2021. https://www.grandviewresearch.com/industry-analysis/3d-printing-industry-analysis.

Manufacturing.gov. "Funding." Accessed January 31, 2022. https://www.manufacturing.gov/funding.

Manufacturing USA. "Institutes." 2022. https://www.manufacturingusa.com/institutes.

Santhanam, Nick, Shekhar Varanasi, Akshay Sethi, and Nidhi Arora. *Value Creation in Industrials*. McKinsey & Company, November 2020. https://www.mckinsey.com/industries/advanced-electronics/our-insights/value-creation-in-industrials.

Sargent, John F. *Manufacturing USA: Advanced Manufacturing Institutes and Network*. Congressional Research Service, March 3, 2021. https://crsreports.congress.gov/product/pdf/R/R46703.

Sohu.com. "What Is KAMP, an Artificial Intelligence Manufacturing Platform Launched in South Korea." January 19, 2021 (in Chinese). https://www.sohu.com/a/445501038_478183.

Stewart, Vivien. "Singapore: Innovation in Technical Education." *Asia Society*. Accessed January 31, 2022. https://asiasociety.org/global-cities-education-network/singapore-innovation-technical-education.

Sublett, Cameron, and David Griffith. "How Aligned Is Career and Technical Education to Local Labor Markets?" The Thomas B. Fordham Institute. April 3, 2019. https://fordhaminstitute.org/national/research/how-aligned-career-and-technical-education-local-labor-markets.

Trex. "Feel Fantastic Recycling Your Plastic! Trex Launches Annual Recycling Challenge for Schools." September 12, 2017. https://www.trex.com/our-company/news

/feel-fantastic-recycling-your-plastic-trex-launches-annual-recycling-challenge-for-schools/.

US Bureau of Labor Statistics. "Industry at a Glance: Food Manufacturing: NAICS 311." Accessed January 31, 2022. https://www.bls.gov/iag/tgs/iag311.htm#workforce.

US Bureau of Labor Statistics. "National Occupational Employment and Wage Estimates." 2020. https://www.bls.gov/oes/current/oes_nat.htm#00-0000.

Chapter 2: The Home Court Advantage

Gartner. "Gartner Says Global Chip Shortage Expected to Persist Until Second Quarter of 2022." May 12, 2021. https://www.gartner.com/en/newsroom/press-releases/2021-05-12-gartner-says-global-chip-shortage-expected-to-persist-until-second-quarter-of-2022.

Klapper, Rebecca. "Low Car Inventory Expected into 2022 as Microchip Shortage Worsens Yet Again." *Newsweek*, September 2, 2021. https://www.newsweek.com/low-car-inventory-expected-2022-microchip-shortage-worsens-yet-again-1625540.

Rapier, Graham. "Elon Musk Says Building a Factory Is '100 Times' as Hard as Building a Car." *Business Insider*, March 15, 2019. https://www.businessinsider.com/elon-musk-says-building-factory-100-times-harder-than-making-car-2019-3.

Seewer, John, and Andrew Welsh-Huggins. "Intel to Build $20B Ohio Chip Facility amid Global Shortage." *Associated Press*, January 21, 2022. https://apnews.com/article/technology-business-lifestyle-games-video-games-cb3e3a36f48416a25b5d00baa405f91a.

Strozewski, Zoe. "Tesla Says No New Models in 2022, Delays Cybertruck Due to Chip Shortage." *Newsweek*, February 9, 2022. https://www.newsweek.com/tesla-says-no-new-models-2022-delays-cybertruck-due-chip-shortage-1677657.

The National Research Council of the National Academies. *Optics and Photonics: Essential Technologies for Our Nation*. 2013. https://doi.org/10.17226/13491.

The ZeroSum Market First Report. "Automotive Inventory Data and Sales Forecasts January 2022." *PR Newswire*, January 26, 2022. https://www.prnewswire.com/news-releases/automotive-inventory-data-and-sales-forecasts-january-2022-301468257.html.

US Department of Commerce. "Commerce Semiconductor Data Confirms Urgent Need for Congress to Pass U.S. Innovation and Competition Act." January 25, 2022. https://www.commerce.gov/news/press-releases/2022/01/commerce-semiconductor-data-confirms-urgent-need-congress-pass-us.

Chapter 3: The Long Haul

Aicardi, Robert. "Clean Harbors CEO: Gulf Cleanup Area 'Like a War Zone.'" *Wicked Local*, June 28, 2010. https://www.wickedlocal.com/story/braintree-forum/2010/06/28/clean-harbors-ceo-gulf-cleanup/39737266007/.

Bandler, James. "The Garbage Czars." *Rutland Herald*, August 23, 1992.

Bulk Handling Systems. "BHS Launches the Max-AI AQC-C." May 1, 2019. https://bulkhandlingsystems.com/wp/bhs-launches-the-max-ai-aqc-c/.

Casella Waste Systems. "History." Accessed January 31, 2022. https://www.casella.com/history.

Clean Harbors. "New Hazardous Waste Incinerator Comes Online." April 4, 2017. https://www.cleanharbors.com/fr/node/2631.

Coudriet, Carter. "How Two Vermont Brothers Built, Trashed, Then Revived Their $1.8 Billion Waste Management Company." *Forbes*, June 28, 2019. https://www.forbes.com/sites/cartercoudriet/2019/06/28/vermont-casella-brothers-waste-systems/?sh=3e61bd986cef.

Davis, Jeremy. "The History of Vermont Skiing: One Hundred Years of Growth." The New England Lost Ski Areas Project, April 20, 1998. https://www.nelsap.org/vt/history.html.

Doler, Kathleen. "John Casella Literally Turns Trash into Treasure over Four Decades." *Investor's Business Daily*, January 7, 2021. https://www.investors.com/news/management/leaders-and-success/casella-waste-systems-ceo-turns-trash-into-treasure-over-four-decades/.

Heyn, Beth. "Clean Harbors Is Featured on 'Undercover Boss.'" *Heavy.com*, January 27, 2020. https://heavy.com/entertainment/2020/01/clean-harbors-undercover-boss/.

National Oceanic and Atmospheric Administration. "Deepwater Horizon Oil Spill: Long-Term Effects on Marine Mammals, Sea Turtles." April 20, 2017. https://oceanservice.noaa.gov/news/apr17/dwh-protected-species.html.

Redling, Adam. "Clean Harbors COO Talks About the Company's Evolution." *Construction & Demolition Recycling*, December 5, 2018. https://www.cdrecycler.com/article/clean-harbors-interview-hazardous-waste/.

Sabataso, Jim. "Stafford, Casella Launch Training Program." *Rutland Herald*, November 9, 2020. https://www.rutlandherald.com/news/covid19/stafford-casella-launch-training-program/article_f06dc245-48f6-54cf-a8e6-23773656aa13.html.

Smith, Howard. "Can This Waste Manager Help Clean Up Your Portfolio?" *Motley Fool*, March 6, 2020. https://www.fool.com/investing/2020/03/06/can-this-waste-manager-help-clean-up-your-portfoli.aspx.

US Bureau of Labor Statistics. "Labor Force Statistics from the Current Population Survey." Accessed January 31, 2022. https://www.bls.gov/cps/data.htm.

Walsh, Molly. "Two Men's Trash: How Casella Waste Systems Converted Garbage into a Sprawling Empire." *Seven Days*, February 19, 2019. https://www.sevendaysvt.com/vermont/two-mens-trash-how-casella-waste-systems-converted-garbage-into-a-sprawling-empire/Content?oid=26608444.

Chapter 4: The Titanium Playbook

Brown, Adam. "The 47,500% Return: Meet the Billionaire Family Behind the Hottest Stock of the Past 30 Years." *Forbes*, January 13, 2020. https://www.forbes.com/sites/abrambrown/2020/01/13/heico-mendelson/?sh=5b1dd8f84b18.

El-Bawab, Nadine. "Here's How Much Americans Have Saved in Their 401(k)s at Every Age." *CNBC*, February 24, 2021. https://www.cnbc.com/2021/02/24/how-much-americans-have-saved-in-their-401k-by-age.html.

HEICO. "Jet Avion: About Us." Accessed January 31, 2022. https://www.heico.com/about-us/subsidiaries/jet/.

National Center for Education Statistics. "Number and Percentage of Public School Students Eligible for Free or Reduced-Price Lunch, by State: Selected Years, 2000–01 Through 2015–16." *Digest of Educational Statistics*, 2017. https://nces.ed.gov/programs/digest/d17/tables/dt17_204.10.asp.

Thermal Structures. "Non-Metallic Fire Shields." Accessed January 31, 2022. https://www.thermalstructures.com/products/non-metallic-fire-shields/.

Vanguard. *How America Saves 2021*. June 2021. https://institutional.vanguard.com/content/dam/inst/vanguard-has/insights-pdfs/21_CIR_HAS21_HAS_FSreport.pdf.

Chapter 5: The Power of Micro-Verticals

BOSS Magazine. "Humble, Hungry, and Smart: The Middleby Food Processing and Bakery Group Is Driven to Perfection." May 2021. https://thebossmagazine.com/hungry-humble-and-smart/.

Foodservice Equipment & Supplies. "The Ultimate Kitchens for Culinary Innovation and Demonstration." May 18, 2021. https://fesmag.com/sponsored/19330-the-ultimate-kitchens-for-culinary-innovation.

Middleby. "Middleby Acquires Ve.Ma.C. Srl." March 26, 2018. https://www.middleby.com/newsroom/middleby-acquires-ve-ma-c-srl/.

Middleby. "Middleby Innovation Kitchens." Accessed January 31, 2022. https://www.middleby.com/mik/.

Middleby. "Middleby Innovation Kitchens Podcast," 2021. https://www.middleby.com/podcasts/middleby-innovation-kitchens/.

Opal Investment Research. "Middleby: Operational Momentum Extends the Multi-Year Growth Runway." *Seeking Alpha*, September 4, 2021. https://seekingalpha.com./article/4453536-middleby-operational-momentum-extends-the-multi-year-growth-runway.

Reuters. "Welbilt Says Ali Group $3.3 Billion Bid Likely Better Than Middleby Deal," May 28, 2021. https://www.reuters.com/article/us-welbilt-m-a-aligroup-idUKKCN2D91UQ.

Root, Al. "This Small-Cap Stock Could Soar as Restaurants Recover." *Barron's*, March 26, 2021. https://www.barrons.com/articles/welbilt-a-small-cap-stock-for-the-restaurant-recovery-51616796000.

The NPD Group. "Consumers Take Grilling and Outdoor Cooking to the Next Level, Reports NPD." April 13, 2021. https://www.npd.com/news/press-releases/2021/consumers-take-grilling-and-outdoor-cooking-to-the-next-level-reports-npd/.

Chapter 6: The Great Amplification Cycle

Cullen, Tim. *Disassembled: A Native Son on Janesville and General Motors—a Story of Grit, Race, Gender, and Wishful Thinking and What It Means for America*. Mineral Point, WI: Little Creek Press, 2019.

Data USA. "About: Simpsonville, SC." 2019. https://datausa.io/profile/geo/simpsonville-sc#about.

Dietrich, Kevin. "S.C.'s Journey to Advanced Manufacturing." *Greenville Business Magazine*, April 30, 2021. https://www.greenvillebusinessmag.com/2021/04/30/354982/s-c-s-journey-to-advanced-manufacturing.

Elias, Jennifer. "Tesla Says It Helped Create More Than 50,000 Jobs in California in 2017." *Silicon Valley Business Journal*, May 16, 2018. https://www.bizjournals.com/sanjose/news/2018/05/16/tesla-economic-impact-california-bay-area-jobs.html.

Goldstein, Amy. *Janesville: An American Story*. New York: Simon & Schuster, 2017.

Goldstein, Amy. "What Is Janesville, Wisconsin, Without General Motors?" *The Atlantic*, April 18, 2017. https://www.theatlantic.com/business/archive/2017/04/janesville-wisconsin-gm-economic-future/523272/.

Greenville County Schools. "U.S. News Announces New Rankings of Elementary and Middle Schools." October 27, 2021. https://www.greenville.k12.sc.us/news/main.asp?titleid=2110usnews.

Jamieson, Claire E. "Change in the Textile Mill Villages of South Carolina's Upstate During the Modern South Era." Master's thesis, University of Tennessee, 2010. https://trace.tennessee.edu/utk_gradthes/635.

Jorge, Jeff, and Erich Bergen. *Supply Chain Resilience: Exploring Nearshoring to Unlock New Frontiers of Strength*. Baker Tilly, December 15, 2020. https://www.bakertilly.com/insights/supply-chain-resilience-exploring-near-shoring-to.

McKinsey & Company. *The Future of Work in America: People and Places, Today and Tomorrow*. July 11, 2019. https://www

.mckinsey.com/featured-insights/future-of-work/the-future-of-work-in-america-people-and-places-today-and-tomorrow.

Michelin North America Inc. "Michelin Ranked in the Top 5 Companies as the Only Tire Manufacturer Among 'Best-in-State Employers' in South Carolina." September 24, 2020. https://michelinmedia.com/pages/blog/detail/article/c0/a1009/.

National Center for Education Statistics. "Number and Percentage of Public School Students Eligible for Free or Reduced-Price Lunch, by State."

Reshoring Initiative. "Reshoring Initiative IH2021 Data Report." September 20, 2021. https://reshorenow.org/blog/reshoring-initiative-ih2021-data-report/.

Reuters. "Welbilt Says Ali Group $3.3 Billion Bid Likely Better Than Middleby Deal."

Simpsonville, South Carolina. "City Again Makes List of Best Places to Live in State." Accessed January 31, 2022. https://www.simpsonville.com/community/page/city-again-makes-list-best-places-live-state.

South Carolina Department of Commerce. "Success Story: The Boeing Company." Accessed January 31, 2022. https://www.sccommerce.com/why-sc/success-stories/boeing-company.

US Bureau of Labor Statistics. "Industry at a Glance: Food Manufacturing: NAICS 311."

Chapter 7: Titanium Family Values

Bailey, Doug. "More Than 8 Out of 10 family businesses Have No Succession Plans." *Boston Globe,* February 9, 2016. http://sponsored.bostonglobe.com/rocklandtrust/more-than-8-out-of-10-family-businesses-have-no-succession-plans/.

Emmett, Jonathan, Asmus Komm, Stefan Moritz, and Friederike Schultz. "This Time It's Personal: Shaping the 'New Possible' Through Employee Experience." McKinsey & Company, September 30, 2021. https://www.mckinsey.com/business-functions/people-and-organizational-performance/our-insights/this-time-its-personal-shaping-the-new-possible-through-employee-experience.

Smith Family Business Initiative at Cornell. "Family Business Facts." Accessed January 31, 2022. https://www.johnson.cornell.edu/smith-family-business-initiative-at-cornell/resources/family-business-facts/.

Varga, Tom. "Ready For Re-D?" *Food Logistics*, May 26, 2010. https://www.foodlogistics.com/transportation/cold-chain/article/10255660/foodservice-distributors-are-taming-their-supply-chain-with-redistribution.

Chapter 8: Winning the Talent War

Amoyaw, May. "Apprenticeship America: An Idea to Reinvent Postsecondary Skills for the Digital Age." *Third Way*, June 11, 2018. https://www.thirdway.org/report/apprenticeship-america-an-idea-to-reinvent-postsecondary-skills-for-the-digital-age.

Apprenticeship.gov. "Registered Apprenticeship Program." 2022. https://www.apprenticeship.gov/employers/registered-apprenticeship-program.

Dixon-Fyle, Sundiatu, Kevin Dolan, Vivian Hunt, and Sara Prince. "Diversity Wins: How Inclusion Matters." McKinsey & Company, May 19, 2020. https://www.mckinsey.com/featured-insights/diversity-and-inclusion/diversity-wins-how-inclusion-matters.

Expatrio. "German Dual Apprenticeship System." Accessed January 31, 2022. https://www.expatrio.com/studying-germany/german-education-system/german-dual-apprenticeship-system.

Farrell, Diana, Martha Laboissiere, Imran Ahmed, Jan Peter aus dem Moore, Tilman Eichstadt, Lucia Fiorito, Alexander Grunewald, et al. "Changing the Fortunes of America's Workforce: A Human-Capital Challenge." McKinsey Global Institute, June 2009. https://www.mckinsey.com/featured-insights/employment-and-growth/changing-the-fortunes-of-us-workforce.

Federal Ministry of Education and Research. "The German Vocational Training System." Accessed January 31, 2022. https://www.bmbf.de/bmbf/en/education/the-german-vocational-training-system/the-german-vocational-training-system_node.html.

Generation USA. "Verizon Partnership." 2020. https://usa.generation.org/partners/verizon/.

Hanson, Melanie. "Average Cost of College & Tuition." *Education Data Initiative*, January 27, 2022. https://educationdata.org/average-cost-of-college.

Hrynowski, Zach. "Nearly Half of U.S. Parents Want More Noncollege Paths." *Gallup*, April 7, 2021. https://news.gallup.com/poll/344201/nearly-half-parents-noncollege-paths.aspx.

Hunt, Vivian, Dennis Layton, and Sara Prince. "Why Diversity Matters." McKinsey & Company, January 1, 2015. https://www.mckinsey.com/business-functions/people-and-organizational-performance/our-insights/why-diversity-matters.

Hunt, Vivian, Lareina Yee, Sara Prince, and Sundiatu Dixon-Fyle. "Delivering Through Diversity." McKinsey & Company, January 18, 2018. https://www.mckinsey.com/business-functions/people-and-organizational-performance/our-insights/delivering-through-diversity.

Ideal Industries. "Rebuilding the Trades," November 11, 2021. Presentation by Meghan Juday at Global Industrial Leadership Summit hosted by McKinsey & Company.

Manyika, James, and Kevin Sneader. "AI, Automation, and the Future of Work: Ten Things to Solve For." McKinsey & Company, June 1, 2018. https://www.mckinsey.com/featured-insights/future-of-work/ai-automation-and-the-future-of-work-ten-things-to-solve-for.

Mishel, Lawrence. "Yes, Manufacturing Still Provides a Pay Advantage, but Staffing Firm Outsourcing Is Eroding It." *Economic Policy Institute*, March 12, 2018. https://www.epi.org/publication/manufacturing-still-provides-a-pay-advantage-but-outsourcing-is-eroding-it/.

National Center on Education and the Economy. "Top-Performing Countries: Singapore." Accessed January 31, 2022. https://ncee.org/country/singapore/.

New York Times. "Job Training That's Free Until You're Hired Is a Blueprint for Biden." April 7, 2021. https://www.nytimes.com/2021/04/07/business/job-training-work.html.

Norheim, Karen. "Female Leadership Has a Ripple Effect in the Manufacturing Industry." *Future of Business and Tech*. Accessed January 31, 2022. https://www.futureofbusinessandtech.com/manufacturing/female-leadership-has-a-ripple-effect-in-the-manufacturing-industry/.

Replogle, Jill. "Where Are All the Construction Tradeswomen?" *Marketplace*, August 17, 2021. https://www.marketplace.org/2021/08/17/where-are-all-the-construction-tradeswomen/.

Sabataso. "Stafford, Casella Launch Training Program."

South Korea Ministry of Education. "High School Vocational Education Advancement Measure." May 12, 2010. http://english.moe.go.kr/boardCnts/view.do?boardID=265&boardSeq=1789&lev=0&searchType=null&statusYN=W&page=18&s=english&m=03&opType=N.

Tucker, Marc S. "The Phoenix: Vocational Education and Training in Singapore." October 2012. https://www.ncee.org/wp-content/uploads/2014/01/The-Phoenix1-7.pdf.

Ydstie, John. "How Germany Wins at Manufacturing—for Now." *NPR*, January 3, 2018. https://www.npr.org/2018/01/03/572901119/how-germany-wins-at-manufacturing-for-now.

Chapter 9: Sustainability Should Be the Last Word

AeroAggregates of North America. (Website). Accessed January 31, 2022. https://www.aeroaggna.com/?gclid=Cj0KCQi

AxoiQBhCRARIsAPsvo-wzVR9_8QWYkyY_4X5N20Q2IkzuBE1rva-Vy5zhCDK3T0uB9NLjUmMaAhBLEALw_wcB.

Blue Planet Systems. "Permanent Carbon Capture." 2021. https://www.blueplanetsystems.com/.

Field, Karen. "John Deere's Tech-Fueled Mission to Feed a Hungry World, One Seed at a Time." *Fierce Electronics*, January 12, 2021. https://www.fierceelectronics.com/electronics/john-deere-s-tech-fueled-mission-to-feed-a-hungry-world-one-seed-at-a-time.

John Deere. "Deere & Company Is Among World's Top 50 Most Admired Companies." February 16, 2017. https://www.deere.com/sub-saharan/en/our-company/news-media/news-releases/2017/february/deere-and-company-among-worlds-top-most-admired-companies/.

Lydon, Bill. "John Deere Digitalization & Automation Increases Farming Productivity." *Automation.com*, February 15, 2021. https://www.automation.com/en-us/articles/february-2021/john-deere-digitalization-automation-farming.

McKinsey & Company. *Climate Math: What a 1.5-Degree Pathway Would Take*. April 30, 2020. https://www.mckinsey.com/business-functions/sustainability/our-insights/climate-math-what-a-1-point-5-degree-pathway-would-take.

NASA. "How NASA and John Deere Helped Tractors Drive Themselves," April 18, 2018. https://www.nasa.gov/feature/directorates/spacetech/spinoff/john_deere.

PwC. "State of Climate Tech 2021." 2021. https://www.pwc.com/gx/en/services/sustainability/publications/state-of-climate-tech.html.

Ramsden, Keegan. "Cement and Concrete: The Environmental Impact." *Princeton Student Climate Initiative*, November 3, 2020. https://psci.princeton.edu/tips/2020/11/3/cement-and-concrete-the-environmental-impact.

Syntax. "Facing Hardware Refresh, Trex Looks to Cloud for JDE|SQL Workloads." February 2020. https://www.syntax.com/wp-content/uploads/2021/02/SYN_SS_Trex_FINAL.pdf.

Trex. "Eco-Friendly, Recycled Plastic Decking." Accessed January 31, 2022. https://www.trex.com/why-trex/eco-friendly-decking/.

United Nations. "Step Up Climate Change Adaptation or Face Serious Human and Economic Damage." *UN Environment Programme*, January 14, 2021. https://www.unep.org/news-and-stories/press-release/step-climate-change-adaptation-or-face-serious-human-and-economic.

United States Environmental Protection Agency. "Inventory of U.S. Greenhouse Gas Emissions and Sinks." 2022. https://www.epa.gov/ghgemissions/inventory-us-greenhouse-gas-emissions-and-sinks.

US Securities and Exchange Commission. "SEC Proposes Rules to Enhance and Standardize Climate-Related Disclosures for Investors." 2022. https://www.sec.gov/news/press-release/2022-46.

Vivint Solar. "How Many Solar Panels Would You Need to Power the USA?" Accessed January 31, 2022. https://www.vivintsolar.com/learning-center/how-many-solar-panels-to-power-the-usa.

World Green Building Council. "Bringing Embodied Carbon Upfront." Accessed January 31, 2022. https://www.worldgbc.org/embodied-carbon.

Zimmer, Anthony, NRMRL, and HakSoo Ha. "Buildings and Infrastructure from a Sustainability Perspective." Environmental Protection Agency. September 2014. https://www.epa.gov/sites/default/files/2016-09/documents/buildingsandinfrastructurefromasustainabilityperspective.pdf.

Chapter 10: The Titanium Disruption

Arora, Nidhi, Nick Santhanam, Shekhar Varanasi, Akshay Sethi. "Value creation in industrials." McKinsey & Company, November 2020. https://www.mckinsey.com/industries/advanced-electronics/our-insights/value-creation-in-industrials.

Bughin, Jacques, Eric Hazan, Susan Lund, Peter Dahlstrom, Anna Wiesinger, and Amresh Subramaniam. "Skill Shift: Automation and the Future of the Workforce." McKinsey Global Institute, May 23, 2018. https://www.mckinsey.com/featured-insights/future-of-work/skill-shift-automation-and-the-future-of-the-workforce.

EOS. "EOS 3D Printing Technology for Aerospace." 2021. https://www.eos.info/en/presscenter/expert-articles/aerospace_liebherr_ventilblock.

Farrell, et al. "Changing the Fortunes of America's Workforce: A Human-Capital Challenge."

Fraunhofer-Gesellschaft. "Fraunhofer-Gesellschaft Profile / Structure."

Gartner. "Gartner Forecasts Worldwide Public Cloud End-User Spending to Grow 23% in 2021." April 21, 2021. https://www.gartner.com/en/newsroom/press-releases/2021-04-21-gartner-forecasts-worldwide-public-cloud-end-user-spending-to-grow-23-percent-in-2021.

Government of Singapore. "About SkillsFuture." Accessed January 31, 2022. https://www.skillsfuture.gov.sg/AboutSkillsFuture.

Grand View Research. *3D Printing Market Size, Share & Trends Analysis Report . . . 2021–2028*. May 2021. https://www.grandviewresearch.com/industry-analysis/3d-printing-industry-analysis.

Jacob, Brian A. "What We Know About Career and Technical Education in High School." *Brookings*, October 5, 2017. https://www.brookings.edu/research/what-we-know-about-career-and-technical-education-in-high-school/.

Lincicome, Scott. "Manufactured Crisis: 'Deindustrialization,' Free Markets, and National Security." Cato Institute, January 27, 2021. https://www.cato.org/publications/policy-analysis/manufactured-crisis-deindustrialization-free-markets-national-security.

Manufacturing USA. "Institutes."

Manufacturing.gov. "Funding."

Manyika, James, Katy George, Eric Chewning, Jonathan Woetzel, and Hans-Werner Kaas. "Building a More Competitive US Manufacturing Sector." McKinsey Global Institute, April 15, 2021. https://www.mckinsey.com/featured-insights/americas/building-a-more-competitive-us-manufacturing-sector.

Polen, Jeff. "The Mask Mover." By Karen Duffin. *NPR*, audio and transcript (April 17, 2020). https://www.npr.org/transcripts/837216447.

Rosalsky, Greg. "Are We Firing Too Many People in the U.S.?" By Karen Duffin. *NPR*, April 7, 2020. https://www.npr.org/sections/money/2020/04/07/828081285/are-we-firing-too-many-people.

Rep. Ryan, Tim, and Tom Reed. H.R.2900—116th Congress (2019–2020): Chief Manufacturing Officer Act (2019). https://www.congress.gov/bill/116th-congress/house-bill/2900.

Sargent. *Manufacturing USA: Advanced Manufacturing Institutes and Network.*

Sohu.com. "What Is KAMP, an Artificial Intelligence Manufacturing Platform Launched in South Korea."

Stewart. "Singapore: Innovation in Technical Education."

Sublett and Griffin. "How Aligned Is Career and Technical Education to Local Labor Markets?"

Trex. "Feel Fantastic Recycling Your Plastic! Trex Launches Annual Recycling Challenge for Schools," September 12, 2017. https://www.trex.com/our-company/news/feel-fantastic-recycling-your-plastic-trex-launches-annual-recycling-challenge-for-schools/.

US Bureau of Labor Statistics. "Industry at a Glance: Food Manufacturing: NAICS 311."

US Bureau of Labor Statistics. "National Occupational Employment and Wage Estimates."

US Department of Labor. "US Department of Labor Announces Available Funding to Promote, Develop, Expand

Registered Apprenticeship in Critical, Non-Traditional Industries." September 10, 2021. https://www.dol.gov/newsroom/releases/eta/eta20210910.

Weisbach, Annette. "Germany is using a familiar weapon to prevent layoffs." *CNBC*, April 3, 2020. https://www.cnbc.com/2020/04/03/kurzarbeit-germany-is-using-a-familiar-weapon-to-prevent-layoffs.html.

Wu, Mark. "The 'China, Inc.' Challenge to Global Trade Governance." *Harvard International Law Journal* 57, no. 2 (spring 2016). https://harvardilj.org/wp-content/uploads/sites/15/HLI210_crop.pdf.

McKinsey & Company

Asutosh Padhi is the managing partner for McKinsey in North America, leading the firm across the United States, Canada, and Mexico and serving as part of McKinsey's fifteen-person global leadership team. He is also a member of McKinsey's Shareholders Council, the firm's equivalent to the board of directors. Working with iconic industrial companies, Asutosh has overseen performance transformations in technology and innovation. He has helped global multinationals bolster their supply chains, reorient strategies, and improve organizational and operational performance. Asutosh is a McKinsey board member and a trustee of the Field Museum in Chicago, one of the top natural history and science museums in the world.

McKinsey & Company

Gaurav Batra previously co-led McKinsey & Company's Advanced Electronics Practice in the Americas. He advised clients on strategy and margin improvement issues, with a particular focus on commercial excellence, pricing, and distributor and channel management. Gaurav also previously led McKinsey's Advanced Data Analytics (ADA) initiative in the Advanced Electronics Practice to help drive sustainable performance improvement for the firm's clients. Prior to joining McKinsey, Gaurav worked as a senior business manager of corporate development with Capital One, a leading diversified financial services company in the United States.

Nick Santhanam is currently the CEO and president of Fernweh Group. He was formerly a senior partner at McKinsey's Palo Alto office and was the leader of the company's global industrials practice. In addition, Nick was the global convener of several McKinsey annual events, including the Industrial CEO event (GILS), the Industrial-Tech and Semiconductors CEO event (T-30), the Food Processing CEO event (FPH-30), and the Senior Tech Influencers Summit (TEDS). He coauthored several of McKinsey's leading industrials research projects, including McKinsey on Industrials, McKinsey on Food Processing and Handling, and McKinsey on Packaging. He is also on the Smithsonian Libraries Advisory Board.

PublicAffairs is a publishing house founded in 1997. It is a tribute to the standards, values, and flair of three persons who have served as mentors to countless reporters, writers, editors, and book people of all kinds, including me.

I. F. Stone, proprietor of *I. F. Stone's Weekly*, combined a commitment to the First Amendment with entrepreneurial zeal and reporting skill and became one of the great independent journalists in American history. At the age of eighty, Izzy published *The Trial of Socrates*, which was a national bestseller. He wrote the book after he taught himself ancient Greek.

Benjamin C. Bradlee was for nearly thirty years the charismatic editorial leader of *The Washington Post*. It was Ben who gave the *Post* the range and courage to pursue such historic issues as Watergate. He supported his reporters with a tenacity that made them fearless and it is no accident that so many became authors of influential, best-selling books.

Robert L. Bernstein, the chief executive of Random House for more than a quarter century, guided one of the nation's premier publishing houses. Bob was personally responsible for many books of political dissent and argument that challenged tyranny around the globe. He is also the founder and longtime chair of Human Rights Watch, one of the most respected human rights organizations in the world.

• • •

For fifty years, the banner of Public Affairs Press was carried by its owner Morris B. Schnapper, who published Gandhi, Nasser, Toynbee, Truman, and about 1,500 other authors. In 1983, Schnapper was described by *The Washington Post* as "a redoubtable gadfly." His legacy will endure in the books to come.

Peter Osnos, *Founder*